Process Development Division 2013

Core Programming Area at the 2013 AIChE Annual Meeting: Global Challenges for Engineering a Sustainable Future

San Francisco, California, USA
3 – 8 November 2013

ISBN: 978-1-63439-052-1

Printed from e-media with permission by:

Curran Associates, Inc.
57 Morehouse Lane
Red Hook, NY 12571

CURRAN ASSOCIATES INC.
proceedings.com

Some format issues inherent in the e-media version may also appear in this print version.

Printed by Curran Associates, Inc. (2014)

AIChE
3 Park Avenue
New York, NY 10016-5991

Phone: (203) 702-7660
Fax: (203) 775-5177

www.aiche.org

Additional copies of this publication are available from:

Curran Associates, Inc.
57 Morehouse Lane
Red Hook, NY 12571 USA
Phone: 845-758-0400
Fax: 845-758-2634
Email: curran@proceedings.com
Web: www.proceedings.com

TABLE OF CONTENTS

(23a) Predicting Surfactant-Related Properties for Chemical-Based Product Design

Mattei, M., Technical University of Denmark
Kontogeorgis, G., Technical University of Denmark
Gani, R., Technical University of Denmark
Originally presented on: 11/3/2013 15:30:00 - 15:55:00

The chemical industry is changing, going beyond commodity chemicals to higher value added products. These are structured products, constituted of numerous chemicals, but still mainly designed and analyzed through experimental techniques. A systematic approach for the design of these products could significantly reduce both time and cost connected to the product development by reducing the amount of experiments needed, provided reliable property prediction methods.

Recently, systematic methodologies for the design of chemical based liquid formulations have been developed, but many of the chemical-based products, especially household and personal care products, are emulsions where active ingredients, surfactants, solvents, additives, *etc.*are mixed together to determine a desired emulsified product. A major difference between the design of liquid formulation and that of emulsion based products is the presence of two immiscible phases, kept together in a stable form by proper emulsifiers, most often surfactants. The presence of these ingredients in the emulsified products is clearly crucial and then their selection is one of the key-steps of the product development. Selection criteria for surfactants consider their effectiveness, safety and toxicity issues, environment, cost or a combination of the aforementioned criteria. Therefore, it is necessary to have available as many properties as possible in order to perform a proper selection based on reliable data. Unfortunately, experimental data are scarce and for some of the key-properties one often finds different values related to different models that cannot be verified due to the absence of an experimental value.

Because of the scarce availability of experimental data, property prediction models for surfactants are needed: critical micelle concentration, cloud point, Krafft temperature and surface tension reduction are considered fundamental properties to have access to in order to perform a proper surfactant design for emulsions. Even though none of these are strictly pure component properties, it is possible to adopt group-contribution methods to describe the available experimental data and then to extrapolate the correlation results to perform predictions. Quantitative structure-property relationship (QSPR) models have already been developed in relation to several of these properties, but in this work it has been decided to use group-contribution methods, based on the Marrero and Gani model, for its easier use in the computer-aided molecular design, predictive power an reliability.

For nonionic surfactants, new third order groups based on the characteristic structures of the most common families of surfactants have been defined and they are included in the Marrero and Gani model. Results from the modeling of critical micelle concentration, cloud point and surface tension reduction are presented and discussed. Also preliminary results regarding the modeling of critical micelle concentration, Krafft temperature and surface tension reduction are presented and discussed for ionic surfactants.

The application of these property models in the systematic methodology for computer-aided product design for emulsion based products is then highlighted.

(23b) Computational Scheme for Rational Solvent Design in Extraction of Pharmaceutical Ingredient

Madakashira, H., Indian Institute of Technology Bombay
Adhikari, J., Indian Institute of Technology, Bombay.
Noronha, S. B., I.I.T. Bombay, Mumbai
K, Y. R., Indian Institute of Chemical Technology
Originally presented on: 11/3/2013 15:55:00 - 16:20:00

The pharmaceutical industry is one of the largest users of organic solvents in processes such as reaction, formulation, separation and cleaning of equipment. Solvents constitute the major part of the mass in manufacture of an active pharmaceutical ingredient (API) with typical E-factors (kg of waste per kg of product) ranging from 25 to 100.[1] An attempt to minimize the amount and number of solvents in a process is essential with regard to cost, environmental, health and toxicological concerns. It is also desirable for these solvents to be green towards reducing their impact on the environment. An experimental trial and error approach to finding an appropriate solvent is time consuming and expensive; computer aided molecular design[2] (CAMD) can be an alternative strategy. CAMD utilizes structure property correlations and optimization technique to find a novel or potential solvent for a physical separation process provided the property estimates are sufficiently accurate. This approach has been applied to a case study where an aliphatic solvent is designed for efficient extraction of an API, R-phenylacetylcarbinol, from an aqueous phase. Current industrial practice involves the use of the solvent toluene.

The prime properties that are of interest for liquid-liquid extraction include distribution coefficient, solvent loss, solvent power, selectivity, viscosity, boiling point and toxicity. These properties have been predicted using group contribution and connectivity index based structure property correlations. The constraints on properties have been framed by choosing toluene as a reference solvent. A variant of genetic algorithm based optimization approach, namely non-dominated sorting genetic algorithm[3] (NSGA-II) along with a UNIFAC[4] (universal functional activity coefficients) subroutine has been employed for optimization. A total of 64 UNIFAC groups (including repetitions) were considered for the current study whose combinations lead to the design of different type of solvent molecules such as aldehydes, ketones, ethers and esters. UNIFAC based structural constraints are employed to design a structurally feasible molecule. A multi objective function minimizing solvent loss and maximizing distribution coefficient have been used to arrive at globally best molecules. The CAMD approach yielded a pareto set of optimal solvents including, amyl acetate and 2-methyl penta 1,3-diene.

Isobaric-isothermal Gibbs ensemble Monte Carlo simulations have been carried out on the solvents identified above, using MCCCS *towhee* package[5]. Simulation results confirm the property estimation predictions and hence, the short-list of solvents chosen for further investigation. Experiments are performed with these solvents for determination of the extent of their suitability for R-PAC recovery.

References:

1. Grodowska, K. & Parczewski, A. Organic solvents in the pharmaceutical industry. *Acta poloniae pharmaceutica* **67**, 3–12 (2010).

2. Harper, P. M. , Gani, R., Kolar, P., Ishikawa, T. Computer-aided molecular design with combined molecular modeling and group contribution. *Fluid Phase Equilibria* **158-160**, 337–347 (1999).

3. Deb, K., Member, A., Pratap, A., Agarwal, S. & Meyarivan, T. A Fast and Elitist Multiobjective Genetic Algorithm. *IEEE transactions on evolutionary computation* **6**, 182–197 (2002).

4. Magnussen, T. UNIFAC Parameter Table for Prediction of Liquid-Liquid Equilibria. *Industrial Engineering Chemistry Process Design and Development* **20**, 331–339 (1981).

5. MCCCS Towhee. Available at: http://towhee.sourceforge.net.

(23c) Molecular Models Using VBA To Determine Components In Petroleum Refining Processes

Vilas, T. G. D., University of Sao Paulo
Oller, C., University of Sao Paulo
Originally presented on: 11/3/2013 16:20:00 - 16:45:00

This work aims to build a database for classifying constituent components of asphaltenes in molecular groups. The classification of these compounds is reached by the similarity of their molecular weights with the use of homologous series groups, using data from the analysis of mass spectrometry and a search algorithm made in VBA language with the goal of identifying a molecule in a specific set of known molecules (database). The method proposed in this work is based on the classification of the compounds through their molecular weights, i.e., the compounds in a mixture are separated into classes by a mass difference previously set. Due to the fact that the database is generated from compounds whose composition is known, it is accepted that a given compound composition has all the possible compositions. To generate molecules based on knowledge of the oil composition, it is used the general formulas of organic compounds which are formed by homologous series. Such series are generated and their molecular weights are calculated and stored together with their molecular formulas, in ascending order of mass, being able to generate the molecular groups with the help of a computing routine in VBA language.

(23d) Understanding Physical Properties of Mixtures of Ionic Liquids

Fillion, J., University of Notre Dame

Xia, H.,

DeSilva, A., University of Notre Dame

QuirozGuzman, M., University of Notre Dame

Brennecke, J. F., University of Notre Dame

Originally presented on: 11/3/2013 16:45:00 - 17:10:00

Ionic liquids (ILs), which are composed of only ions, are liquids with a melting point less than 100 °C. In addition, most ILs have negligible vapor pressure. The anion and cation can both be adjusted, allowing for millions of different ILs to exist. Due to the enormous amount of potential ILs that can be made, trends need to be discovered in order to tune the ILs properly to fit the needs of various industrial applications. One such application is CO_2-IL co-fluid vapor compression refrigeration technology, which requires a certain enthalpy of reaction between the IL and the CO_2 and a low viscosity for the IL. The vapor compression cycle involves CO_2 and IL to be pump around a desorber, compressor, resorber, and expansion device. The benefit of CO_2 as refrigerant instead of HFC-134a is that CO_2 has a Global Warming Potential, GWP, of 1, instead of a GWP of 1410. Of course, the GWP of ILs is essentially 0 because of their low volatility. The temperature and pressure range will vary in the cycle, from 10 °C to 50 °C, and from 10 bars to 30 bars. Through proper tuning, ILs have been created to meet the desired enthalpy. However, the viscosity of these ILs are too high at lower temperatures. Such a high viscosity would cause a huge cost to pump the ILs throughout the cycle.

Understanding the properties of ILs and how the properties change once the ILs are adjusted is important. Also, examining how ILs properties change when another IL or a diluent (e.g., tetraglyme) is added will offer another means to obtain the desired properties. Mixtures of ILs are important because it offers the option to use desirable benefits from two different ILs. In this work, the primarily focus for the mixture properties is the viscosity. In addition, we present several mixture densities and CO_2 solubilities. The mixture glass transition, T_g, the melting point, T_m, and the decomposition, T_{onset}, are measured using Differential Scanning Calorimetry, DSC, and Thermal Gravimetric Analysis, TGA,.

The mixture densities always fall between the densities of the pure ILs or the IL and the tetraglyme. By contrast, the mixture viscosities exhibit a wide range of behaviors. Some mixtures have viscosities that smoothly transition between the viscosities of the two pure compounds. However, other mixtures exhibit viscosities higher than either of the pure ILs. In general, mixtures tend to exhibit eutectic behavior, where the melting points of the mixtures are lower than the melting point of either pure IL. However, in many instances, the addition of a second IL will eliminate a melting point, leading to a glass transition temperature at much lower temperatures. The trends in these behaviors, in terms of alkyl chain length, fluorination, ether chains and ammonium or phosphonium cation, will be discussed.

(23e) Ionic Liquid Electrolytes for Lithium Ion Batteries-Effect of Lithium Salts On Thermal and Transport Properties

Wu, L., Clarkson University
Venkatanarayanan, R. I., Clarkson University
Lebga-Nebane, J. L., Clarkson University
Shi, X., Clarkson University
McLaughlin, J. B., Clarkson University
Roy, D., Clarkson University
Krishnan, S., Clarkson University
Originally presented on: 11/3/2013 17:10:00 - 17:35:00

Ionic liquids (ILs) are of significant interest as non-volatile and non-flammable electrolytes for lithium ion batteries.[1] Towards developing more stable and more conductive Li-ion battery electrolytes, the thermal and transport properties of the solutions of two different lithium salts (lithium iodide and lithium bistriflamide) in six different ILs [1,3-dialkylimidazolium iodide,[2] 1,3-dialkylimidazolium bistriflamide, PEGylated imidazolium iodide,[2] PEGylated imidazolium bistriflamide,[3] (ω-perfluoroalkyl)PEG-derivatized imidazolium iodide,[3] and (ω-perfluoroalkyl)PEG-derivatized imidazolium bistriflamide[3]] were characterized using differential scanning calorimetry, viscometry, electrochemical impedance spectroscopy, and pulsed field gradient nuclear magnetic resonance diffusion experiments. PEGylation improved lithium salt solubility in the ILs. Complexation of lithium ions with the ether oxygen atoms of the PEG moieties strongly affected thermal properties, such as the glass transition temperature, and transport properties such as viscosity, self-diffusion coefficient, and ionic conductivity. The properties of the IL–salt blends were analyzed using a recently reported mathematical framework relating viscosity, diffusion coefficient, and conductivity.[4] Structure–property correlations were also investigated through molecular dynamics simulations conducted using Materials Studio.

References

1) Rock, S.; Lin, W.; Crain, D.; Krishnan, S.; Roy, D. Interfacial Characteristics of a PEGylated Imidazoliujm Bistriflamide Ionic Liquid Electrolyte at a Lithium Ion Battery Cathode of $LiMn_2O_4$. *ACS Appl. Mater. Interfaces* **2013**, *5*, 2075-2084.

2) Ganapatibhotla, L. V. N. R.; Zheng, J.; Roy, D.; Krishnan, S. PEGylated Imidazolium Iodide Ionic Liquid Electrolytes: Thermophysical and Electrochemical Properties. *Chem. Mater.* **2010**, *22*, 6347-6360.

3) Ganapatibhotla, L. V. N. R.; Wu, L.; Zheng, J.; Jia, X.; Roy, D.; McLaughlin, J. B.; Krishnan, S. Ionic Liquids with Fluorinated Block-oligomer Tails: Impact of Self-assembly on Transport Properties. *J. Mater. Chem.* **2011**, *21*, 19275-19285.

4) Lebga-Nebane, J. L; Rock, S.; Franclemont, J.; Roy, D.; Krishnan, S. Thermophysical Properties and Proton Transport Mechanisms of Trialkylammonium and 1-Alkyl-1*H*-imidazol-3-ium Protic Ionic Liquids. *Ind. Eng. Chem. Res.* **2012**, *51*, 14084-14098.

(23f) Use of Induction Heating As An Inexpensive Process to Dramatically Improve the Structure and Grain Size of Compound Semiconductor Absorber Layers for Photovoltaic Applications

Chojecki, G. S., Clarkson University
Rasmussen, D. H., Clarkson University
Originally presented on: 11/3/2013 17:35:00 - 18:00:00

The performance of thin-film photovoltaic devices is highly dependent on the microstructural quality of the absorber layer. The high defect densities characteristic of $Cu(In,Ga)Se_2$ (CIGS) and $Cu_2ZnSn(S,Se)_4$ (CZTSSe) thin-films lead to reductions in the minority carrier lifetime, mobility and even the optical absorption of the resulting material. This has led to the development of controlled growth methods to produce dense thin-film absorber layers exhibiting large, highly-crystalline grains. However, these processes are typically expensive, increasing the ever-important performance metric, cost per watt. Here, we report the use of induction heating under an applied pressure as an inexpensive, novel process for the synthesis of high quality semiconductor thin-films from particulate precursor materials. The process has been applied to CIGS, CZTSSe and other complex chalcogenide absorber layers.

(37a) The Carbon Capture Simulation Initiative (CCSI): Computational Tools To Accelerate Commercial Development

Miller, D. C., U.S. Department Of Energy, National Energy Technology Laboratory

Originally presented on: 11/4/2013 8:30:00 - 8:55:00

The U.S. Department of Energy's Carbon Capture Simulation Initiative (CCSI) is developing, demonstrating and deploying advanced computational tools and models to accelerate the development of next generation technologies, specifically the development of cost effective carbon capture technologies. This CCSI Toolset will (1) enable promising concepts to be evaluated more quickly based on an optimized process, (2) reduce the time for design and troubleshooting by integrating process and device scale simulations to better predict performance and more effectively resolve issues, (3) quantify technical risk to enable more focused scale up activities, and finally, (4) stabilize the costs during commercial deployment by having a more comprehensive understanding of the process and underlying behavior of the system. This paper will provide an overview of the program and highlight recent accomplishments applying the tools to potential carbon capture technologies.

(37b) The Four Dimensional Product: Integration Over Time Is the Only Way to Understand Sustainability

Banholzer, W., Dow Chemical Company
Jones, M. E., The Dow Chemical Company
Originally presented on: 11/4/2013 8:55:00 - 9:20:00

Sustainability is a question of time, integrating knowledge of production, use and end-of-life. Far too often, hype has surpassed reality, boundary conditions have been neglected and actions taken that don't truly lead to a sustainable future. Singular focus on attributes like renewable content or biodegradability gives the wrong answer. Quality engineering forms the foundation for any meaningful discussion of sustainability, providing insights into energy and material flows, energy return, and thermodynamic constraints. The chemical industry is increasing resource and energy efficiency through innovation in both products and processes. Our mastery of materials science enables unique energy producing and energy saving products, products that improve sustainability by being significantly better than the next best alternatives during use.

(37c) Maximization of Energy Efficiency of a Combined Heat and Power Plant

Edgar, T. F., The University of Texas at Austin
Powell, K., The University of Texas at Austin
Kim, J., The University of Texas at Austin
Kapoor, K., The University of Texas at Austin
Originally presented on: 11/4/2013 9:20:00 - 9:45:00

Energy efficiency is receiving increased attention as a way to reduce the use of fossil fuels and the resulting production of greenhouse gases. Automation, process control, and real-time optimization are key technologies to operate plants in the most efficient way. A combined heat and power plant at the University of Texas at Austin has been studied in order to reduce its carbon footprint as well as reduce costs. The steam power plant has been modeled using steady state mass and energy balances. Chillers have also been modeled so that their efficiencies can be maximized. There is also a thermal storage tank for chilled water to be used during periods of high demand. Forecasting of loads as well as the effect of weather have been incorporated into the real-time optimization approach. The effects of time of day pricing of power and use of demand response techniques to flatten demand curves are discussed. Usage of thermal energy storage along with automatic control gives additional degrees of freedom to deal with dynamic power conditions.

(37d) Pan American Issues on Biofuel Sustainability

Shonnard, D. R., Michigan Technological University

Originally presented on: 11/4/2013 9:45:00 - 10:10:00

Biofuels are in rapid development in many Pan American countries, yet the nature and extent of impacts on social and environmental systems from large-scale biofuel development are not well understood. This talk will explore sustainability issues across a broad set of impacts and in different Pan American countries and will begin to outline important research gaps. A future research agenda to overcome these knowledge gaps would help establish a new and growing biofuels industry onto a sustainable path.

(62a) Session Honoring Don Woods

Faucher, S., Hatch
Keith, J. M., Michigan Technological University
Originally presented on: 11/4/2013 8:30:00 - 10:35:00

Session Honoring Don Woods

Abstract

This session honors Professor Don Woods who passed away in April 2013. Dr. Woods was a chemical engineering professor at McMaster University from 1964 to 2000 where he used innovative teaching methods and won many teaching awards (and three honorary doctorates from Queens, Guelph and McMaster). He wrote many books and papers, and gave workshops in various parts of the world. What you may not know about Dr. Woods is that he was also an artist, banjo player, square dancer, badminton player, builder, author, genealogist, and great reader. He enjoyed spending time at Kinlough, the family's 200 acre vacation place in southern Bruce County, as well as travelling to many interesting destinations in retirement.

This session will highlight reflections on the impact of Dr. Woods career. Speakers in this session include:

- Angie Perna, NJIT

- Phil Wood, McMaster University

- Rich Felder, North Carolina State University

- Scott Fogler, University of Michigan

- Charles Petty, Michigan State University

- Suzanne Kresta, University of Alberta

- George Chase, The University of Akron

(62b) Cross Flow Filtration Hydrocyclones

Petty, C. A., Michigan State University

Originally presented on: 11/4/2013 10:35:00 - 11:00:00

A cross flow filtration hydrocyclone (CFFH) combines the desirable attributes of a vortex separator and a filter barrier. By selecting an appropriate hydrophilic filter medium, a CFFH clarifier has the potential to produce a filtrate phase with a low concentration of a dispersed organic phase in a single stage. The CFFH concept addresses three traditional problems associated with current hydrocyclone separators for oil/water separation: 1) the loss of separation performance due to core flow reversal; 2) the loss of separation performance due to entrained oil in the sidewall boundary layer; and, 3) the loss of separation performance due to hydraulic turndown. The paper will discuss the efficacy of using a forward-flow CFFH and a reverse-flow CFFH for clarifying produced water on offshore platforms and elsewhere. Computational fluid dynamic (CFD-) methods are used to identify practical prototype designs and operating characteristics.

"The Ubiquitous Hydrocyclone: Theory and Practice, C. A. Petty, AIChE Webinar, February 11, 2009, http://www.aiche.org/webinars.aspx.

C. A. Petty and S. M. Parks , 2001, "Flow Predictions within Hydrocyclones", Filtration+Separation, Volume 38, Number 6, 28-34.

Ali, S.K., K. Weispfennig, J.F. Foss, and C. A. Petty, 1992, "The Separation Performance of Long Hydrocyclones with an Annular Feed Entry", BHRA 4th Int'l Conf. on Hydrocyclones, Southampton, UK, 23-25 September 1992.

Dabir, B. and C. A. Petty, 1986, "Measurements of Mean Velocity Profiles in a Hydrocyclone Using Laser Doppler Anemometry", Chem. Eng. Commun., 48, 377-388.

Bahram Dabir, 1983, "Mean Velocity Measurements in a 3"-Hydrocyclone Using Laser Doppler Anemometry, Michigan State University, Ph.D. Dissertation, 1983.

Knowles, S. R., D.R. Woods, and I.A. Feinstein, 1973, The Velocity Distribution within a Hydrocyclone Operating without an Air Core", CAN J CHEM ENG, 51, no.3, pp.263-271.

(77a) A Commercial Demonstration of Biorefinery of Lipids----Coproduction of Biodiesel and 1,3-Propanediol

Liu, D., Tsinghua University
Originally presented on: 11/4/2013 8:30:00 - 9:00:00

Lipase-catalyzed transesterification (also esterification) from renewable oils for biodiesel production has some advantages over chemical-catalyzed transesterification, such as environmental friendship, lower energy consumption□widely suitable for low quality vegetable oil. However, the low stability due to alcohol and the expensive cost of the lipase had been being regarded as the main hurdle to the industrialization of lipase-catalyzed biodiesel production. Tsinghua University has proposed a novel process to eliminate the inhibition of methanol on lipase, thus the operational life of the immobilized lipase could be improved into more than 300 cycles from several (1-5) cycles at the traditional enzymatic applied process. After the successful demonstration in a pilot plant with capacity of 200kg/day biodiesel, the first commercial facilities with capacity of 30,000ton/year was constructed in Hunan Rivers Bioengineering Co., Ltd, China and it was put into operation on Dec. 8, 2006. On the basis of the promising results of demonstration at a pilot-plan in Brazil, the Brazilian ministry of science and technology decided to support the commercialization of such a new technology in Brazil.

As a by-product, glycerol will be yielded at about 10% of biodiesel during the process of biodiesel production. More and more concerns are pay on how to utilize glycerol. It could be a promising way to produce 1,3-propanediol (PDO) by fermentation with crude glycerol as feedstock. PDO is a valuable chemical product and especially it could be copolymerized with terephthalic acid (or methyl ester) to form polytrimethylene terephthalate (PTT). PTT has excellent properties for textile fibre or plastic, and it was found to have some better properties than other polyesters such as PET, PBT. Tsinghua University has developed a novel flexible process for PDO production by fermentation from glycerol or glucose, and the demonstration was finished in pilot plant at the end of 2003. A facility with capacity of 4,000 tons/year PDO is being run in Hunan Rivers Bioengineering Co. Ltd, China.

(77b) Accelerated Process Development and Scale Up

am Ende, D.,

Originally presented on: 11/4/2013 9:00:00 - 9:30:00

One of the goals of process development for chemical manufacture (either for Pharmaceuticals, Agrochemicals, Fine-chemicals, or Energetics etc) generally involves identification of an economical and viable route to the target compound, followed by scale-up efforts to manufacture quantities of the desired product. Leveraging the right tools can dramatically reduce the time and resources needed to successfully achieve these goals. Several case studies and applications highlighting important time-saving tools and approaches through process development will be presented.

(77c) Using Dimensional Analysis to Commercialize Products and Processes

Worstell, J. H., Worstell and Worstell, Consultants

Originally presented on: 11/4/2013 9:30:00 - 10:00:00

The key to successful product commercialization is successful process scaleup. Originally, in the chemical process industry (CPI), scaleup meant building successively larger blackboxes; i.e., pilot plants, into which inputs went and products emeged. Then, with the advent of analog and finally digital computers, the hope arose of turning scaleup into an exercise involving the numerical solution of the pertinent conservation laws. Chemical engineers have pursued this calculational effort since the mid-1960's in order to reduce the size and number of blackboxes between product conceptualization and product commercialization. The push since the mid-1980's for inherently safe pilot plants has also contributed to the CPI's effort to reduce the number and size of blackboxes built during a given product commercialization. Unfortunately, the mathematical equations describing most chemical processes are inordinately complex and difficult to solve, even numerically. Thus, chemical engineers need to develop new methods for scaling products and processes from conceptualization to commercialization. This presentation describes and discusses using Dimensional Analysis to reduce the number and size of pilot plants built during such a commercialization and it discusses how Dimensional Analysis can reduce the size and scope of the experimental program done in a pilot plant.

(77d) Process Intensification; Challenge, Targets and Experience

Siebenhofer, M., Graz University of Technology
Lux, S., Graz University of Technology
Aksamija, E., Graz University of Technology
Feiner, R., Graz University of Technology
Originally presented on: 11/4/2013 10:00:00 - 10:30:00

In teaching chemical engineering the Eastman Kodak process serves as a welcome demonstration for process intensification in methyl acetate synthesis. History and development of this specific process features several aspects of process intensify-cation in chemical engineering, covering the technology, process design, hydraulics as well as economics. Although dating back to the eighties this process is still a benchmark, worth matching with, in particular and in general. Beside facing the challenge of direct process related competition we may also draw valuable conclu-sions for future guidelines in new fields of applications from the history, the design strategy and the outcome of this demonstration example.

Do we identify any future applications in the pipeline? When talking about future applications, these challenges have already emerged on a selfenhancing and selfaccelerating global scale. Following the recommendations of several biorefinery roadmaps we expect from biorefinery a span of contributions and new technologies to several aspects of energy management, product and process development as well as sustainable effects on a safer environment. And, we may address the complete toolbox of process intensification. The basic question is how to get process intensification in biorefinery technologies started and successfully applied? By identifying thermodynamic limits and technological obstacles, the specification of process needs as well as product related needs is an approach we may simply base project initiation on, since we talk about multi feedstock multiproduct mixtures with complex interactions due to the nature and chemistry of substances.

The synthesis of methyl acetate is a welcome basic project. Synthesis of methyl acetate is limited by the chemical equilibrium, azeotropic product mixtures as well as low relative volatility of binary systems. The basic industrial scale Eastman Kodak process with successful application of process intensification, has been mentioned. Does it therefore leave any driving force for further intensification activities? It does, since implementation of pervaporation offers an alternative technological approach for synthesis and separation and product isolation.

Separation of azeotropic mixtures is a very representative example for applying the process intensification toolbox in biorefinery. Exemplarily the distillation of formic acid/acetic acid/water-mixtures, frequently formed in biomass processing, is more or less impossible, since we are faced with a saddle azeotrope. Reactive separations give access to a variety of process concepts which help overcome the thermodynamic limits of separation.

When addressing separations in biorefinery applications, liquid/liquid-extraction processes may suffer from third phase formation and emulsification. Measures in design and construction may help overcome operation limits.

Application of process intensification tools in development and design of future biofuel production is a high priority task and must, if we expect biofuels to contribute significantly to fueling traffic and industry in future, be trementously intensified. Seemingly the transfer of Fischer-Tropsch technology from coal conversion to application in biomass conversion is expected to serve as state of the art technology, but overall process efficiency and yield limitation strongly recommend intensified investigation of alternative routes to finally obtain fuels with sufficient calorific value and stability at an acceptable oxygen level originating in the feed.

With this practical experience in mind the discussion of rules, strategies and recommendations in process development, design and intensification is a welcome opportunity to check the content of the toolbox and the effectiveness of tools.

(77e) Toward Faster Commercialization of New Materials

Musso, C., McKinsey & Company
Originally presented on: 11/4/2013 10:30:00 - 11:00:00

Successful commercialization of new materials is notoriously difficult, often taking 10+ years to recover development investments. While materials may never reach the commercialization speeds of many consumer products, they can be significantly improved. This presentation will expose the pitfalls of commercialization, along with practical suggestions for avoiding them, showing practices to shorten the commercialization time of new materials by up to 50%. It will draw on both rigorous analysis of historical plastics and lessons learned from the development dozens of launch strategies.

Bio of presenter: Christopher Musso is a principal in McKinsey's Chemicals and Product Development practices. He leads the Firm's work on commercialization of new materials, and has done ~20 commercialization projects for leading chemicals players, ranging from bio-based plastics to new commodity variants to high-end specialty plastics. He specializes in developing market insight for new applications and creating new business models to maximize penetration while minimizing commercialization time. He holds a PhD in Technology, Management, and Policy from MIT, where he studied the commercialization patterns of plastic and developed a model that has underpinned much of his work at McKinsey. In addition to his PhD, Chris holds Master's degrees in Materials Science, Business Administration, and Technology Policy, all from MIT, where he was a Leaders for Manufacturing Fellow. He earned his Bachelor's degree in Manufacturing from Brigham Young University. He has published several papers on materials engineering and innovation in both popular and refereed journals.

(115a) Succeeding At Transferring Complex Technologies: Partnering Across Organizations

Stamato, H., Bristol-Myers Squibb
Nikfar, F., Bristol-Myers Squibb
Deshpande, S.,
Koo, O., Bristol-Myers Squibb Company
Parab, P.,
Patel, C., Bristol-Myers Squibb Company
Jia, J.,
Yan, B.,
Alves, L., Bristol-Myers Squibb
Scheer, B.,
Marin, A.,
Sidhu, P.,
Rabideau, M., Bristol-Myers Squibb
Peragine, J., Bristol-Myers Squibb
Heeter, V.,
Freed, L.,
Varia, S.,
Timmins, P., Bristol Myers Squibb
Originally presented on: 11/4/2013 12:30:00 - 12:55:00

Technology transfer, while rooted in the flow of materials and creation of documents, is also founded on the communication and relationships forged among organizations both internal and external to the enterprise. Manufacture of pharmaceutical dosage forms involves many different technical disciplines, all with a unique perspective and a contribution of the culture necessary to support routine use of the technology at the receiving site. The more complex the dosage form; the more complex, and risky, the transfer can be. As a regulated activity the requirements and documentation to meet health authority standards must also be considered. The presentation will discuss the various perspectives, their communication flow and how the overall transfer process can be orchestrated to a successful conclusion, as derived from the experiences in technology transfer of complex dosage forms both internally and externally.

(115b) Challenges and Best Practices in the Commercialization of Polycarbonate Copolymers

Khadilkar, M., SABIC Innovative Plastics
Hollar, W., SAUDI BASIC INDUSTRIES CORPORATION
Originally presented on: 11/4/2013 12:55:00 - 13:20:00

The Innovative Plastics SBU of SABIC has successfully commercialized multiple polycarbonate co-polymer products over the past decade. As with any scaleup and commercialization activity, there can be formidable challenges in successfully translating new products/processes to the manufacturing scale. The framework of Design for Six Sigma (DFSS) incorporates many best practices which mitigate challenges including:

- Effectiveness of cross-functional team interactions including communications
- Scaleup risk management
- Project management and control
- Translation to the manufacturing plant

In this presentation the use of selected DFSS tools will be reviewed with examples of how they contribute to successful commercialization.

SABIC is a trademark of Saudi Basic Industries Corporation or its affiliates

(115c) A Strategy for Tablet Active Film Coating Formulation Development Using a Content Uniformity Model and Quality-By-Design Principles

Wang, J., Bristol-Myers Squibb Comany

Originally presented on: 11/4/2013 13:20:00 - 13:45:00

Tablet active film coating is becoming a common approach to overcome some formulation stability and/or manufacturing process challenges for fixed dose combinations of oral solid dosage forms in pharmaceutical industry. If a drug molecule is prone to acid or alkali based degradation, one of the commercially viable formulation approaches is to enhance the chemical stability of a drug molecule by incorporating the active pharmaceutical ingredient in the coating layer of the film coated tablets. The aqueous active film-coating has been shown to be an effective strategy to minimize chemical interactions between different drug molecules by physically separating one active compound from other core tablet components during the development of a fixed dose combination for tablets. There are three common challenges associated with an aqueous active film-coating process. They are: (1) to accurately determine the active coating end-point to consistently achieve the target potency; (2) to achieve satisfactory and consistent tablet content uniformity; and (3) to maximize the amount of active deposited during the coating process. The mechanistic understanding through engineering modeling and thorough control of the coating operation not only address all these issues satisfactorily, but also provide tools to a model-based scale-up to commercial production scale.

(115d) Pathway to Process Technology Commercialization

Euhus, D. D., Sabic Americas, Inc.
Sciamanna, D. S., PolyDiamond Technologies
Bryan, P. F., Chevron Technology Ventures LLC
Originally presented on: 11/4/2013 13:45:00 - 14:10:00

In the refining and chemicals industry, commercialization of new process technology is not nearly as common as it was in days gone by. Because of this, the chemical engineering skill set and institutional memory required to "scale-up" a new technology has been forgotten. This paper presents a pathway for the chemical/process engineer to follow when commercializing a new process technology. From defining and understanding the different scales at which process development occurs to looking at specific steps in the pathway of taking your breakthrough process concept to a commercial production scale, the proposed steps guide the engineer who has not commercialized a new process technology through the stages required to achieve techno-economic success. The risks attenuated by each stage of the pathway are also discussed. These guidelines help the developer of a new process technology to understand the trade-offs between risk and cost before deciding to skip a step of process technology development.

(115e) Polyphenylene Ether Macromonomers

Peters, E. N.,
Carrillo, A., SABIC Innovative Plastics
Fisher, S., SABIC Innovative Plastics
Originally presented on: 11/4/2013 14:10:00 - 14:35:00

PPE has been used for over 20 years to enhance the performance of epoxy based Printed Wiring Boards. However, its use and performance were limited by process complications, limited solubility in solvents and high viscosities. The continuous progression of electronics toward high speed digital, wireless, and the transfer of large data files have placed increased demands on the performance of dielectric materials. An understanding of market needs, the customer's infrastructure and processes, and the ability to manufacture macromonomers in our current assets has led to the successful commercialization of PPE engineering macromonomers. The PPE family of macromonomers offer high solubility in solvents, broad performance enhancements such as lower dielectric properties, lower moisture uptake, higher heat, and improve FR performance. Indeed, these macromers have been heralded as a breakthrough in the search for materials that broadly enhance the performance of dielectric materials.

(115f) Concept to Commercialization - How Can We Improve The Process?

Pace, J. E. III, SABIC

Originally presented on: 11/4/2013 14:35:00 - 15:00:00

The journey toward commercialization of chemical processes is fraught with issues that can reduce the quality of the desired end product – an operating process effectively producing a targeted product to the required specifications and economic performance. As the chemical industry becomes increasingly competitive on a global scale, it is becoming more imperative than ever to avoid these pitfalls. This increasingly competitive environment also dictates that there is a significant advantage for those who can bring processes to commercialization in the shortest possible time. Increased speed of development can lead to more of the issues that can reduce the product and economic quality of the final process, so it is more important than ever to improve the work processes for development and commercialization that allow all of these objectives to be fully met. Methods are discussed which can help reduce project timelines and, at the same time, deliver commercial processes which meet all of their expectations.

(117a) Development of a Subnormothermic Machine Perfusion System for the Human Liver

Bruinsma, B., Center for Engineering in Medicine, Massachusetts General Hospital, Harvard Medical School, Shriners Hospital for Children
Farmer, A.,
Ozer, S., Center for Engineering in Medicine (CEM) at Massachusetts General Hospital â€“ Harvard Medical School - Shriners Hospital for Children
Berendsen, T., MGH/Harvard Medical School
Izamis, M. L., Harvard Medical School / Massachusetts General Hospital
Martins, P.,
Yeh, H.,
Uygun, K., Harvard Medical School
Originally presented on: 11/4/2013 12:30:00 - 12:54:00

The use of livers donated after cardiac death (DCD) of the donor is limited by the increased incidence of primary nonfunction and biliary complications following transplantation of these grafts. While the transplantation field continues to struggle with an ongoing shortage of donor organs, various strategies are being investigated to increase the availability of livers for transplantation, including the exploration of novel techniques to make DCD livers a safer option. Ex vivo machine perfusion holds the potential to strongly increase the quality of a donor liver and improving post-transplantation function and reducing complications. Collaborations with the New England Organ Bank (NEOB), the MGH transplant center and UMass Memorial medical center have provided us with human livers, discarded due to extended warm ischemic time, donor age, excessive steatosis or other contraindications for transplantation. Thus far, we have successfully perfused 8 discarded human livers, showing feasibility of subnormothermic machine perfusion in preserving the human liver ex vivo with active metabolism and demonstrating recovery of viability expressed in significantly improved energy status. Ongoing inclusion of human livers will allow characterization of these livers by broadening the ex vivo liver function testing, to include markers of injury and function.

Isolated (sub)normothermic organ perfusion is challenged by the absence of other homeostatic organs that are involved with maintenance of acid-base physiology (kidney, lung), energy metabolism (pancreas, intestine), fluid and electrolyte balance (kidney). These functions need to be compensated for by exogenous administration of hormones, nutrients, electrolytes and other substrates for metabolism along with a perfusion solution that supports this regulation as an artificial extracellular compartment. Controlling these functions requires a thorough understanding of both hepatic and systemic physiology and a system that supports it by allowing for real-time measurement of various parameters, easy intervention and reliable viability testing.

The subnormothermic system for the perfusion of human organs that we have developed is user friendly, but also enables advanced monitoring of hepatic function and intervention. An artificial cardiopulmonary system, including a set of pumps and hollow fiber oxygenators, drives the flow of an oxygen rich perfusion solution through the portal vein and hepatic artery of the liver. A carbogen mixture of 95% O_2/5% CO_2 creates an inflow pCO_2 of ±75 mmHg, which, in combination with exogenous bicarbonate plays a role in the carbonate buffer system. The pH of the system is periodically monitored and can be controlled by boluses of bicarbonate. Superphysiological partial oxygen pressures of >700 mmHg ensure adequate oxygenation of the tissue in the absence of hemoglobin. A commonly used hepatocyte culture medium is used a base solution for perfusion. Williams medium E is rich in nutrients and antioxidants. Supplemented with hydrocortisone and insulin it has been shown to provide a rich environment supporting hepatocyte viability in ex vivo liver perfusion and increasing survival after transplantation in the rat. Ports integrated into the system allow drawing samples for blood gas analysis, which provides periodical and on demand information about acid/base, oxygenation and electrolyte balance. Transducers proximal to both the portal and arterial inflow of the liver provide live pressure readings that can be used to calculate hepatic resistance. Samples from the perfusion solution can be analyzed for a multitude of metabolites.

Our results show that the human liver can be preserved for a minimum of three hours ex vivo in a subnormothermic setting (±21 °C), with continued metabolic function. A continuous consumption of O_2 and CO_2 output reflects active oxidative metabolism. An initially acidic perfusion solution and high lactate levels are normalized by the end of perfusion and electrolyte levels remain normal. An active biliary system is reflected in the steady production of bile. The portal venous and hepatic arterial circulation decrease in resistance, suggesting an improving circulation. The recovery of energy status of the graft is demonstrated by an increase in ATP content during perfusion. To our knowledge this is the first report of the application of subnormothermic machine perfusion in human livers and this work will provide a basis for use of this system in a clinical trial for liver transplantation.

(117b) Hemodialysis: A Sea Change Needed, Not Realized

Leonard, E. F., Columbia University
Cortell, S., St Luke's/Roosevelt Hospital

Originally presented on: 11/4/2013 12:54:00 - 13:18:00

The treatment of end-stage renal disease, ESRD, in the second decade of the 21st century involves a half-million patients in the US, for whom about $30 billion are spent each year, with fewer than 5% significantly rehabilitated. Surprisingly, after years of exotic research, the accepted most critical problem of these patients is water management, and there is not yet a way to match water removal and water generation rates, which leads to unacceptable accumulations between thrice-weekly treatments A thoroughly researched, 'simple' method for removing water continuously would be a superb sea change, but it doesn't yet work. This paper will review technical approaches, problems, and prospects for therapeutic improvement in a growing patient population surrounded by a cost-contained healthcare system. In particular the promise and new problems involved in ambulatory blood processing will be enumerated and assessed in terms of those that are fundamental, like coagulation and membrane fouling versus those that should respond to known technical innovations. All blood processing devices are nominally microfluidic but biological and artificial microfluidic systems differ in numerous important ways that will also be assessed.

(117c) An Automated Tissue Digester for Pancreatic Islet Production

Lu, Z., Auburn University
Boland, E., Techshot Inc
Todd, P. W., Space Hardware Optimization Technology, Inc
Hanley, T. R., Auburn University
Originally presented on: 11/4/2013 13:18:00 - 13:42:00

Pancreatic digestion is the primary step in preparing islets of Langerhans for clinical transplantation for restoring euglycemia in Type-1 Diabetes patients. In 1988, Ricordi et al. developed a semi-automated method in which pancreas fragments and glass balls were placed into a cylindrical vessel to form a closed-loop recirculation and dissociation system, operated by intermittent manual shaking of the vessel. However, the yield of islets from this technique tends to be variable and depends on several parameters. Collagenase-containing enzyme blends used for releasing islets from the exocrine tissue also produce variable results and can cause damage to islets. Operator function (shaking the vessel by hand) also produces variability from case to case. An automated digestion unit capable of producing reproducible operating parameters is needed for the consistent preparation of islets.

This research focuses on the incorporation of an automated horizontally rotating digester (MDR) consisting of a rotating outer cylindrical shell and a counter-rotating core both with hemispherical baffles. This configuration is designed to enhance the turbulent effect and contact area between tissue fragments and walls. In the digestion process, the rotation of the inner core is adjusted for optimum dissociation, and independently the exterior wall rotation is optimized to maintain particles in suspension without centrifuging them. Rotation speeds are also adjusted as particle size distribution is diminished during digestion. In addition, a discrete phase model in FLUENT is used to simulate the flow patterns and track the tissue particles in the MDR and Ricordi chamber. The computational fluid dynamics (CFD) analysis and comparison of flow variable contours indicate that MDR has predictable, controllable fluid shear environment and optimized differential rotation mode at 120 ml/min velocity inlet and 100 rpm rotation speed in opposite directions. The agreement between experimental and CFD predicted results is shown by analysis of viable islets under a Grasshopper® digital camera with an ultra-fast 800Mb/s digital interface. Finally, the validation of the model is performed through the comparison of simulation results with experimental data obtained from porcine pancreatic tissue digestion.

(117d) Improving the Post-Thaw Wash Process for Frozen Red Blood Cells: An Integrated Theoretical and Experimental Approach

Lusianti, R. E., Oregon State University
Higgins, A. Z., Oregon State University
Originally presented on: 11/4/2013 13:42:00 - 14:06:00

Transfusion of red blood cells (RBCs) is one of the most commonly practiced medical procedures, with approximately 15 million units transfused per year in the United States alone. Currently, blood banks and hospitals use RBCs refrigerated at 4°C in their reserve almost exclusively. Depending on the preservative, the FDA allows a storage period of 21 to 42 days for refrigerated blood. The short storage period makes it difficult to maintain a steady supply of refrigerated blood to meet the unpredictable demand for transfusion. On the contrary, freezing blood in the presence of 40% glycerol extends the FDA approved shelf life of the product to 10 years. The use of frozen blood in blood banking would enable stockpiling strategic reserves to eliminate seasonal shortages, improve availability in the event of a natural disaster, facilitate military use during the time of war, and provide overall stability in the RBC supply chain. Although logistically more advantageous, the use frozen blood in clinical therapy is uncommon due to the time consuming post-thaw wash process required to remove the glycerol before the product can be transfused. The current state of the art removal method requires approximately one hour to process a single blood unit, making short notice use of frozen blood particularly challenging. Until a more rapid method of glycerol removal is discovered, the increased use of frozen blood in blood banking remains unlikely.

In this study, modern approaches such as mathematical modeling and microfluidics were used to develop a method that expeditiously removes glycerol from thawed frozen RBCs without causing excessive cellular damage. A cell membrane transport model was used in guiding the design of several procedures for varying the extracellular solution concentration with time to theoretically remove glycerol in a rapid manner without causing cell membrane damage. Batch experiments employing a series of dilutions were then conducted to test the theoretical deglycerolization procedures, resulting in the identification of a four-step procedure capable of removing glycerol while achieving a cell recovery that meets transfusion standards in a total process time of three minutes. Direct scaled up implementation of the batch procedure would require excessive amounts of diluting solution and produce an extremely dilute cell suspension. Hence, a membrane microfluidic device was built to practically adapt the batch procedure in a continuous manner. Additionally, a model capable of predicting the glycerol removal in the membrane device was developed to assist in selecting the appropriate operational conditions to match the extracellular concentration changes in the batch procedure. Preliminary experiments using thawed frozen blood and 3.4% saline solution in a single device resulted in a reduction of glycerol mass composition of up to 55.0%±0.3% with a corresponding hemolysis of 15.1%±0.7%. The glycerol removal achieved in this experiment is comparable with the first step of the batch procedure, demonstrating partial implementation in a continuous platform. Matching the step-by-step concentration changes from the remaining steps is possible with multiple membrane devices linked together.

The operating conditions necessary to achieve such concentration changes can be predicted using the mathematical model for each device in the assembly. The use of a combination approach of mathematical modeling and microfluidics shows promise in enabling full continuous implementation of the rapid batch deglycerolization procedure. Ultimately, the findings from this study has the potential to dramatically reduce the post processing time of frozen blood, making it a more attractive option for blood banking in the future.

(117e) Polymeric Nanosensors for in Vivo Lithium Monitoring

Cash, K. J., Northeastern University
Clark, H. A., Northeastern University
Originally presented on: 11/4/2013 14:06:00 - 14:30:00

Recent years have shown the incredible versatility and utility of polymeric nanosensors. Sensors have been developed for a wide variety of ionic analytes (e.g. sodium, potassium, pH) as well as nonionic analytics (e.g. glucose) and have been applied for challenging research fields such as detection of sodium sparks in live cardiomyocytes. This field of research has yielded new knowledge on cellular metabolism and function. Additionally, these sensors can be applied in vivo, through implantation in or under the skin in animals; showing potential as a research tool for continuous monitoring of analyte concentrations without the need for direct samples to be taken from the animal. I will present research on the development and testing of *in vivo* lithium nanosensors based on this platform which are able to continuously monitor lithium concentrations for eventual application in the treatment of bipolar disorder.

This work was supported by the Research reported in this publication was supported by the National Institute Of Biomedical Imaging And Bioengineering of the National Institutes of Health under Award Number F32 EB015270, the National Institute of General Medicine of the National Institutes of Health under award number R01 GM084366 and the National Institute of Neurological Disorders and Stroke under award number R01 NS081641.

(117f) Modeling the Ability of Lubricin to Reduce Bacterial Proliferation On the Surfaces of Medical Devices

Ravi, V.,
Aninwene, G. II, Northeastern University
Webster, T. J., Northeastern University
Originally presented on: 11/4/2013 14:30:00 - 14:54:00

Abstract:

Infection of medical devices is a major clinical issue. This study used mathematical modeling to describe and quantify the ability of lubricin, an anti-adhesive glycoprotein found in synovial fluid, to inhibit the proliferation of bacteria on the surfaces of medical devices. In a previous experiment, bacterial growth trials were performed with *Staphylococcus aureus* and *Staphylococcus epidermidis* on tissue cultures with either lubricin, vitronectin, or mucin to investigate how lubricin and protein sub-regions of lubricin may reduce bacterial adhesion and proliferation. Mathematical models were created with the resulting data and were used to describe the growth dynamics of these bacteria in cultures with each of these proteins. The effect on lag time (λ), maximum specific growth rate (μ_m), and the asymptote reached during the stationary phase (A) were determined and compared. Lubricin was found to significantly increase λ of *S. aureus* and *S. epidermidis* and significantly decreased both the μ_m of *S. aureus* and A of *S. aureus* and *S. epidermidis*.

Introduction:

The attachment and proliferation of bacteria on the surface of medical devices threatens the success of these devices. It can lead to a reduction or loss of device function as well as infection, inflammation, and/or damage to the surrounding tissue[1]. *Staphylococcus aureus* and *Staphylococcus epidermidis* colonize a large percentage of the human population[2] and are two of the major culprits of biofouling, the proliferation of unwanted biological material on the surfaces of an implanted device. Following attachment to surfaces, they can form biofilms which are resistant to antimicrobials and difficult for the host's immune system to combat[3].

This study examined the ability of lubricin (LUB), a glycoprotein found in the synovial fluid that has lubricating and anti-adhesive properties, to decrease bacteria functions. LUB has a mucin-like center domain and vitronectin-like tail domains[4]. It is thought that LUB can serve as an effective non-immune opsonification agent for resisting bacteria colonization[5]. If bacteria are unable to colonize a surface and form a biofilm, it will be much easier for the immune system and antimicrobial agents to clear the bacteria before they can cause an infection.

The ability of LUB to inhibit bacterial proliferation is quantified in this study by determining the effect that LUB and protein sub-regions of LUB have on the growth curve of each bacterium. This curve is generally separated into several phases including a lag phase, during which the bacteria are adapting to the conditions of their environment and not dividing, an exponential phase during which exponential growth is seen, and a stationary phase during which the growth and death rate are equal[6]. The quantities determined by our modeling were the lag time (the length of the lag phase), the maximum specific growth rate (the greatest growth rate during the acceleration of the exponential phase), and the asymptote (a value directly correlated to the number of organisms in the stationary phase where the growth and death rates are in equilibrium). These values were determined by fitting the growth data to two sigmoid growth functions: logistic and Gompertz.

Methods:

In a previous experiment, 96-well polystyrene plates were treated with either lubricin (LUB), bovine submaxillary mucin (BSM), or bovine vitronectin (VTN). BSM was used to assess the effect of the central mucin-like domain in LUB, and VTN to assess the effect of the globular vitronectin-like tail domains[4]. LUB used in this study was extracted from bovine synovial fluid (Pel-Freez)[7]. BSM and VTN were obtained from Sigma Aldrich. LUB and BSM were used at a concentration of 200μg/mL. VTN was used at a concentration of 50μg/mL. 50μL of each protein solution were dried on the polystyrene well surface overnight.

Staphylococcus aureus (ATCC (25923)) and *Staphylococcus epidermidis* (ATCC (35984)) obtained from the ATCC (25923) were cultured in tryptic soy broth (TSB) (Sigma Aldrich) for 18 hours to reach stationary phase and were then diluted to a density of 10^7 bacteria/mL. The bacterial cell culture procedure involved inoculating approximately 3 mL of sterile TSB inoculated with one colony of desired bacteria, then incubating the solution on a shaker set to 200 rpm inside an incubator maintained at 37°C. Bacteria were seeded into the wells and were allowed to incubate for 15 minutes in a stationary incubator maintained at 37°C. After 15 minutes the bacterial solution was removed, the plates were rinsed with sterile PBS, and the wells were filled with 200μL of fresh TSB. Optical density measurements were taken every 4 minutes for 24 hours.

For this study, the growth data (optical density vs time) obtained from these bacterial trials were fit to modified forms of the logistic model and the Gompertz model using MATLAB. The model equations used were modified from their generic forms to contain coefficients for the three bacterial growth parameters lag time (λ), maximum specific growth rate (μ_m), and asymptote of the stationary phase (A) by deriving formulas to represent the mathematical parameters in terms of the bacterial growth parameters[8]. The goodness-of-fit was determined using R-square and sum of squares for error (SSE) and the models with the best fit were used to determine the value of the three parameters.

Results and Discussion:

The Gompertz and logistic models both fit the bacterial growth data very closely. The Gompertz model had a slightly better fit for 8 out of 10 data sets, based on the R-square and SSE.

S. aureus: LUB, BSM, and VTN all significantly increased the duration of the lag phase (λ) compared to the PBS soaked control. On average, LUB showed a 27% increase (52 minutes), VTN showed a 43% increase (82 minutes), and BSM showed a 29% increase (55 minutes). LUB and BSM decreased the maximum specific growth rate (μ_m) while VTN increased it. VTN and LUB also decreased the upper asymptote (A). The lag time increases seen with LUB, BSM, and VTN indicate that these proteins were making it more difficult for *S. aureus* to adapt itself to the environment and start growing exponentially. The decrease in μ_m seen with LUB and BSM shows that these proteins are reducing the acceleration of the growth phase. Though the acceleration was reduced, BSM reached a similar asymptote in the stationary phase as the PBS soaked control, while LUB also led to a much larger reduction in the asymptote. Interestingly, VTN increased the μ_m, but still led to a decrease in the asymptote. It was suggested that vitronectin can enhance adhesion in some bacteria[9], which may explain the increase in μ_m.

S. epidermidis: LUB and BSM significantly increased the lag time while VTN decreased it compared to the PBS soaked control. There was a 35% increase (105 minutes) in lag time with LUB, a 30% increase with BSM (89 minutes), and a 2% decrease (6 minutes) with VTN. LUB and BSM increased the maximum specific growth rate while VTN decreased it and increased the asymptote while VTN decreased it. Though the increase in asymptote caused by LUB is statistically significant, the difference is so small as to not indicate a biological or clinical significance. Once again, a significant increase in lag time is seen with both LUB and BSM indicating that these proteins inhibited *S. epidermidis* from adapting to the environment and slowed the entry into the exponential phase.

Conclusions:

Lubricin significantly inhibited the ability of both *S. aureus* and *S. epidermidis* to attach and proliferate on surfaces. The most striking finding was the large increase in lag time. This is particularly significant because it may give the immune system and antimicrobials more time to clear the bacteria before an infection is established.

References:

[1] a) S. D. Puckett, E. Taylor, T. Raimondo and T. J. Webster, *Biomaterials* **2010**, *31*, 706-713; b) D. J. Weber, K. L. Hoffman, R. A. Thoft and A. S. Baker, *Review of Infectious Diseases* **1986**, *8*, 12-20.

[2] S. R. Gill, D. E. Fouts, G. L. Archer, E. F. Mongodin, R. T. DeBoy, J. Ravel, I. T. Paulsen, J. F. Kolonay, L. Brinkac, M. Beanan, R. J. Dodson, S. C. Daugherty, R. Madupu, S. V. Angiuoli, A. S. Durkin, D. H. Haft, J. Vamathevan, H. Khouri, T. Utterback, C. Lee, G. Dimitrov, L. Jiang, H. Qin, J. Weidman, K. Tran, K. Kang, I. R. Hance, K. E. Nelson and C. M. Fraser, *J. Bacteriol.* **2005**, *187*, 2426-2438.

[3] J. W. Costerton, P. S. Stewart and E. P. Greenberg, *Science* **1999**, *284*, 1318-1322.

[4] D. P. Chang, N. I. Abu-Lail, F. Guilak, G. D. Jay and S. Zauscher, *Langmuir* **2008**, *24*, 1183-1193.

[5] G. E. I. Aninwene, E. Taylor, D. Hall, A. Mei, G. D. Jay and T. J. Webster, *MRS Online Proceedings Library* **2012**, *1417.*

[6] M. L. Shuler and F. Kargi, *Bioprocess Engineering: Basic Concepts (Second Edition)*, Prentice Hall PTR, Upper Saddle River, NJ, **2002**, p.

[7] G. D. Jay, D. A. Harris and C.-J. Cha, *Glycoconjugate Journal* **2001**, *18*, 807-815.

[8] M. H. Zwietering, I. Jongenburger, F. M. Rombouts and K. Vantriet, *Applied and Environmental Microbiology* **1990**, *56*, 1875-1881.

[9] B. Singh, Y. C. Su and K. Riesbeck, *Mol Microbiol* **2010**, *78*, 545-560.

(182a) Economic and Engineering Assessment of Combined Heat and Power Fuel Cell Systems Installed in Commercial Buildings

Brooks, K., Pacific Northwest National Laboratory
Pilli, S.,
Originally presented on: 11/4/2013 15:15:00 - 15:37:00

PNNL support of the DOE's Fuel Cells Technology Office is to identify near term niche markets where fuel cells have potential, work with the DOE and stakeholders to develop activities in those areas, analyze the business case, and present the results to the community. Accelerating fuel cell use in these key early markets will create jobs in an industry that needs high volume purchases to ramp up production, decrease costs, support commercialization, and enable a domestic supplier base. PNNL has contracted ClearEdge Power to demonstrate fifteen 5 kW combined heat and power fuel cells in a wide variety of business in California and Oregon. These fuel cells are being remotely monitored for such things as electrical and thermal energy, overall efficiency, and availability. The gathered information will be analyzed, to provide "real-world" data from units "in the customer's hands" to validate performance, durability and reliability; installation, operations, and maintenance costs; and identify remaining barriers to widespread commercialization. As part of this assessment, a business case has been developed estimating of the size of the market and its growth potential, identifying possible niche markets, and comparing the CHP FCS with its alternatives in terms of economics, engineering and environmental impact. The results of this assessment will be presented.

(182b) Systematic Approach for Energy Efficient Eco-Industrial Parks

Noureldin, M. B., Saudi Aramco
Qahtani, S. A., Saudi Aramco
Owaidah, M., Saudi Aramco
Nutaifi, A., Saudi Aramco
Originally presented on: 11/4/2013 15:37:00 - 15:59:00

Sustainability since its original conceptualization is considered to have four dimensions; social; economic; environmental and institutional. It is simply the improved management of natural resources within a business establishment to provide economic and social benefits to the business and its surroundings. Eco-industrial parks (EIP) can serve a significant role in realizing economic, environmental and social benefits both to individual plants and to network of companies. Even though EIP are often publicized as a mean of reducing environmental damage through reduced waste, they can have number of benefits at different levels. Currently the major driving factor for the creation of most of it is to attain financial gain.

Industrial ecology concept may promote a new path of local development through the transition from industrial clusters to EIPs via exploiting common features of both models. Implementing Eco-Industrial Parks' (EIPs) principles in an existing industrial cluster or in the planning of new ones, represents a significant opportunity for its revitalization. IEP could exploit synergies from industrial clusters and non-industrial activities to easily create new production models in which the economic and environmental dimensions are symbiotic.

The well planning of new EIP or transformation of conventional industrial complexes to a one can brings significant value to energy efficiency and energy-based GHG emissions reduction. This task is a huge multi-variable multi-dimensional optimization problem in which the total eco-industrial park network depends on as small as stream condition and as big as the whole park functionality. The customized integration among many industrial plants and non-industrial ones in adjacent geographical locations can bring in more degrees of freedom to the "waste energy recovery" optimization problem and consequently presents new horizon to the radical energy-based GHG emissions reduction to levels never thought of before. Fortunately, many of the reasons which were hindering the application of inter-systems integration are nowadays addressed in cost-effective ways. However, to date EIPs key success has been a sequence of independent economically driven actions. Such evolutionary pattern may not be easily transferred to others. A holistic approach in addressing the problem using systematic techniques in the implementing of necessary modifications in either the brownfields or in the planning of Greenfields can bring huge value to energy resources conservation and GHG emissions reduction and is desperately needed today. It is very difficult with the current methods and tools to magically manufacture EIPs to work from scratch. First, there should be the basic ingredients in place, namely the desire of plants to actively participate and second the correct EIP membership and structure of firms. These basic ingredients can then be enhanced and improved upon, with correct support structure in place. The key to developing a successful EIP is to determine the best connections among different industrial plants and its surrounding communities in an EIP. Specifically it is the material and energy flows relationship among the different members in the alliance of plants which permit establishing optimal linkage to form a fruitful inter-dynamic structure. If such structure does not exist a successful EIP cannot be realized.

This paper addresses the energy efficiency of EIP via new systematic approach. It is different in concept than the current approach in which only the "transfer of waste heat" from one firm to a nearby one is conducted after exhausting all the possibilities to optimize the stand-alone plant in the firm. The total site targeting based on the pinch technology, which is the most widely used to date, starts with intra-systems integration which means again that only the waste heat of one member in the EIP is shared with other parks' members. Certainly, such level of sharing philosophy is not the ideal proactive form of true cooperation.

This paper introduces new method that achieves ambitious levels of energy efficiency and GHG emissions reduction via identifying best matchings' among EIP members; generating technically viable energy efficient alternatives; and defining the best generation and allocation of energy utilities systems. It also presents two new mathematical programming and constraints logic propagation tools to systematically find the best and the second best and so on matchings' among several plants for optimal eco-industrial park membership forms under all possible combinations of plants-specific design and operations modifications. The paper also introduces new graphical tool that enables the EIP designer to gain insights and create initial solutions for optimal energy loads for exchange among the members to best enhance the energy efficiency and GHG emissions of the whole park ("One for all and all for one philosophy"). The advantages of the new method and tools are illustrated and demonstrated by two case studies.

(182c) The Potential Of Osmotic Membrane Dehumidification

Kesten, A. S., Nanocap Technologies LLC
McCutcheon, J. R., University of Connecticut
Girelli, A.,
Blechner, J. N., Nanocap Technologies LLC
Originally presented on: 11/4/2013 15:59:00 - 16:21:00

THE POTENTIAL OF OSMOTIC MEMBRANE DEHUMIDIFICATION

Arthur Kesten, Jeffrey McCutcheon, Ariel Girelli and Jack Blechner

An osmotic membrane dehumidifier can use a flexible, semi-permeable membrane to facilitate capillary condensation of water vapor and the transport of condensed water through the membrane into a salt solution by osmosis. Here a humid gas stream is brought into contact with a semi-permeable membrane, which separates the gas stream from an osmotic (e.g., salt) solution. Some of the pores of the membrane are small enough to permit capillary condensation. Liquid formed within these pores can connect with liquid formed in adjacent pores, collectively forming continuous paths of liquid. These ?liquid bridges' extend across the thickness of the semi-permeable membrane and provide paths by which water can travel across the membrane. Because the membrane is so thin, water concentration gradients across the membrane can be large. This can provide a large driving force for water transport between the humid air and the osmotic fluid. The flexibility of the polymeric membrane allows for considerable design flexibility that enhances the potential for retrofit with any cooling system. An illustration of this two-step spontaneous process is given below.

Liquid desiccants are particularly effective as osmotic agents because water entering the desiccant solution is bound to the salt as water of hydration. This enhances the water concentration gradient across the membrane.

Laboratory testing of membranes and draw solutions under different environmental conditions is conducted using a cell comprised of two halves separated by a membrane. On the top half, the humid air is passed above the membrane. The draw solution is pumped through the bottom half of the cell; Â the draw solution is contained in a reservoir that is placed on a digital scale that measures mass to a hundredth of a gram. Humid air contacting the membrane condenses by capillary condensation in the pores of the membrane and the condensed water is drawn by osmosis into the draw solution. Mass changes versus time are recorded with the digital scale that is capable of measuring a maximum of 4,000 grams.

Osmotic dehumidification begins with condensation of water molecules, followed by the osmotic draw of the liquid water out of the pores into a draw solution. If the rate-limiting step is the osmotic de-swelling of the membrane, the highest osmotic pressure should be used to maximize flux. However, if capillary condensation is the limiting step, a threshold is reached beyond which increased osmotic pressure does not increase flux. To understand these limits, we tested three concentrations of magnesium chloride with a cellulose acetate membrane under identical conditions. The water flux varied directly with concentration suggesting that the limiting step for this membrane was the osmotic removal of liquid water as opposed to capillary condensation.

A separate experiment was constructed to compare the effectiveness of <u>no</u> membrane to our capillary condensation/osmotic dehydration system. Having no membrane in the system is the equivalent of exposing liquid desiccant directly to a humid air stream. Measured water removal rates for that system were less than one fifth of the rates for the membrane/draw solution tested here. And, of course, with no membrane between the humid air and the desiccant, there is always the potential for entraining desiccant in the air stream.

Cooling of the osmotic solution results in more effective dehumidification. Lower water vapor pressure leads to capillary condensation at lower relative humidity. For modest levels of cooling, the relative humidity exiting the Nanocap process will be around 50%.

Scale up of the process to larger surface areas of membrane can be accomplished with plate and frame systems, spiral wound devices and bundles of cylindrical fibers. Results with a plate and frame system demonstrate the effectiveness of a mathematical model as a basis for design of a larger system.

OSMOTIC DEHYDRATION

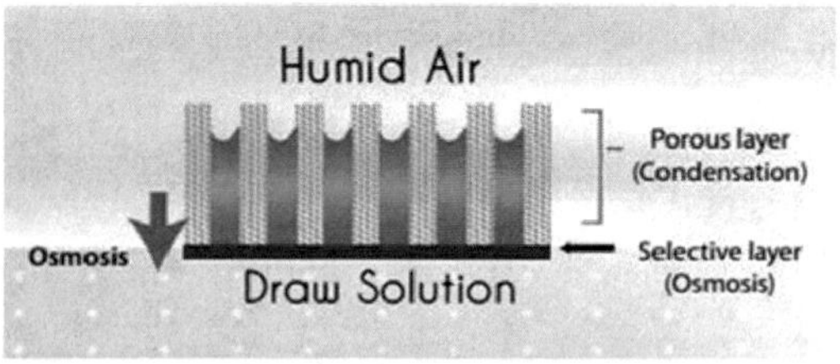

Osmotic dehydration across an anisotropic membrane consisting of a selective layer that is supported by a porous layer. Capillary condensation allows for osmosis to occur across the selective layer of the membrane into a concentrated salt or osmotic solution.

THE NANOCAP PROCESS

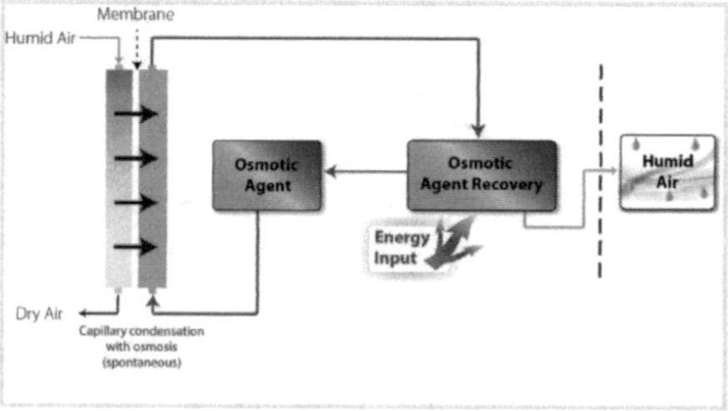

Energy used to regenerate the osmotic agent is intended to be a solar or waste heat source. Regeneration of the osmotic agent can be continuous or intermittent; the optimum time to regenerate occurs when solar or reject heat is available.

(182d) Improvement of Distillation Column Efficiency By Integration With Organic Rankine Power Generation Cycle

Sladkovskiy, D. A., St.Petersburg State Institute of Technology (technical university)
Originally presented on: 11/4/2013 16:21:00 - 16:43:00

Distillation as main separation technology consumes greater than 40% of the energy used by the chemical and petroleum process industries but the energy efficiency of a distillation column is low. It is well-known that significant energy savings can be achieved if complex column configurations and heat integration are used, although these improvements withdraw the attractive advantages of conventional distillation column such as flexibility, low capital investment and low operational risk.

This theoretical research focuses on improvement of distillation column vapor generation procedure by expansion turbine installation.

The liquid stream from the column bottom is pressurized by a pump and the high pressure liquid is vaporized in the reboiler. The resultant high pressure vapor stream drives the expansion turbine, which is coupled to the electricity generator or another rotating equipments. The turbine exhaust vapor enters the column bottom.

Such distillation column with power generation cycle (DCPGC) is kind of combination of conventional distillation column and organic Rankine cycle (ORC) power plant. The ORC is a cost efficient and proven method of converting low temperature waste heat to mechanical and/or electrical energy.

Such chemical components and its mixtures as propane, i-butane, n-butane, i-pentane et al. which occurs in distillation columns bottoms are known to be utilized or considered as ORC working fluids. So DCPGC is seem to be practically feasible.

DCPGC reboiler duty is about 10-20% greater than the conventional column reboiler duty but almost whole of extra heat transforms to electricity power. The theoretical DCPGC electricity generation efficiency is more than 90% (the heat required for the distillation remains the same as in conventional column and is not included in the efficiency calculation).

Only thermodynamic properties and thermal stability of column's bottom product impact on DCPGC feasibility, but not the temperature difference between top and bottom of the column. This is an advantage of suggested technology to compare with heat integrated column with heat pumps where temperature difference is quite important.

This theoretical research describes practical and economical feasibility of DCPGC for separating propane/propylene, i-butane/n-butane, i-pentane/n-pentane, butane-pentane, benzene/toluene, xylene mixtures. Crude oil and naphtha fractionation are also investigated. For instance DCPGC for i-butane/n-butane binary mixtures separation with 25 tonne/h feed flowrate can produce 1 MW of electricity power.

(182e) High Surface Area Zeolite Membrane for Efficient Bio-Fuel Production

LIU, W., Pacific Northwest National Laboratory

Originally presented on: 11/4/2013 16:43:00 - 17:05:00

Distillation and adsorption has been commonly used for respective enrichment and polishing of ethanol fuels in today's ethanol plants. In this presentation, development of thin zeolite/metal sheet membranes of surface area packing density comparable to polymeric sheets is discussed to enhance ethanol dehydration energy efficiency and lower capital costs. The mechanical structural integrity of this new membrane product concept is substantiated with stress modeling. A relatively simple fabrication process that renders its low-cost manufacturing is delineated with experimental results. Separation process studies are conducted along with the membrane product development, which include (i) performance tests with real-world feeds and (ii) process design and simulation of membrane separation integrated with whole plants. The present membrane is highly permeable to water that significant one-pass dehydration can be obtained within a short residence time (seconds). However, the membrane performance can be drastically affected by the feedstock. Proper pre-treatment is necessary to achieve stable separation performances with the real-world feed. Based on experimental results, a two-stage membrane separation system is proposed to conduct ethanol enrichment and purification by replacing respective distillation and adsorption processes. More than 90% energy saving is shown using the membrane separation. Due to high permeance and high membrane area packing density, the membrane module estimated for a 40-million gal/year ethanol plant can be fairly compact, about the size of plate heat exchangers. This membrane technology is promising for retrofitting of existing corn ethanol plants and for building of new cellulosic ethanol plants in the future as well.

(182f) Syngas Mass Transfer Analysis in a Hollow Fiber Reactor

Orgill, J. J., Brigham Young University
Lewis, R. S., Brigham Young University
Atiyeh, H. K., Oklahoma State University
Originally presented on: 11/4/2013 17:05:00 - 17:27:00

Biofuel production via fermentation is produced primarily by fermentation of simple sugars. Besides the sugar fermentation route, there exists a promising alternative process that uses syngas (CO, H_2, CO_2) produced from biomass as building blocks for biofuels. Although syngas fermentation has many benefits, there are several challenges that still need to be addressed in order for syngas fermentation to become a viable process for producing biofuels on a large scale. One challenge is mass transfer limitations due to low solubilities of syngas species. The hollow fiber reactor (HFR) is one type of reactor that has the potential for achieving appropriate mass transfer rates for biofuels production, but a better understanding of mass transfer limitations associated with all syngas species is critical for assessing syngas usage during fermentation.

Mass transfer coefficients for O_2 and CO_2 have been typically assessed for HFRs using dissolved gas probes. In limited cases, CO has been studied using myoglobin assays. Though there are many reports on mass transfer coefficients in HFRs, studies are lacking that compare mass transfer coefficients of syngas species, including comparisons obtained from different measurement methods. Such comparisons are important to verify the accuracy of the experimental methods and to provide an increased understanding of the competing mass transfer limitations associated with syngas species.

In this study, we measured the mass transfer coefficient for O_2 in a PDMS hollow fiber reactor as a function of Reynolds number (flow rate) using a dissolved O_2 probe. The data was fit to a model characterizing O_2 mass transfer and the model was used to predict the mass transfer coefficients for CO and H_2. In order to verify the model predictions for these latter species, a second measurement technique using GC analysis was developed to measure CO and H_2 mass transfer coefficients. The GC technique was validated by comparing the mass transfer coefficient of O_2 from both GC and dissolved O_2 probe analysis. The mass transfer coefficients of both H_2 and CO in a HFR will be compared and the impact of such coefficients on syngas utilization will be presented. This study will greatly enhance the knowledge necessary to assess syngas usage and mass transfer limitations in a HFR used for syngas fermentation.

(182g) An Innovative Methanol Synthesis Process Using Self-Heat Recuperation

Kansha, Y., The University of Tokyo
Ishizuka, M., The University of Tokyo
Tsutsumi, A., The University of Tokyo
Originally presented on: 11/4/2013 17:27:00 - 17:49:00

The demand for methanol will continue increasing, since methanol is attracting as a fuel for fuel cells, and is an intermediate raw material of hydrogen and dimethyl ether (DME) which are categorized green energy source. As well as other chemical processes, many methanol synthesis processes have reactant recycle system with product separation due to the low conversion ratio of the reactor. For this product separation from reactant recycle stream, a gas-liquid separator or distillation process has often been used. However, this gas-liquid separator and distillation processes are well known as an energy consuming process. Thus, many chemical engineers and investigators managed to produce a catalyst which achieves high conversion from reactant to production in the reactor for whole process optimization.

Authors have developed self-heat recuperation technology, in which not only the latent heat but also the sensible heat of the process stream can be circulated without any heat addition, leading to reduction of energy requirement in several chemical processes. This self-heat recuperation technology is suitable for thermal and separation processes and is applied to several chemical processes for energy saving as case studies.

In this research, we investigated the feasibility of applying self-heat recuperation technology to the methanol synthesis process and developed an innovative process for methanol synthesis process from the energy saving point of view. By installing the self-heat recuperation technology to the methanol synthesis process, the energy consumption of the process can be greatly reduced. Thus, it can be said that installation of self-heat recuperation technology is the other option of process design method to achieve high conversion in the reactor.

(220c) Novel Approach in Heat Exchanger Fouling Monitoring

Erdener Akinc, H., TUPRAS
Becer, M., TUPRAS
Orman, S., TUPRAS
Sahin, G., TUPRAS
Bakir, M., TUPRAS
Bas Hekimoglu, S., TUPRAS

<align="center">Novel Approach in Heat Exchanger Fouling Monitoring

H. Erdener Akinç[1], M. Becer[2], G. Sahin[2], S. Orman[1], S. Bas Hekimoglu[1], M. Bakir[1]

[1]TÜPRAS, Turkish Petroleum Refineries Corporation Head Office, Kocaeli/TURKEY

[2]TÜPRAS, Turkish Petroleum Refineries Corporation Izmir Refinery, Kocaeli/TURKEY

(Corresponding Author: metin.becer@tupras.com.tr)

Energy management is an important aspect for every industry especially for refineries. Due to everchanging refinery margins; the prevention of energy losses and maximum energy recovery are hot topics. Fouling in preheat exchangers is considered as the main sources of energy loss, reaching up to 2% of the refinerys' total energy consumption. It is of great concern due to lowering of heat exchanger capacities and thus causing an increase in furnace heat loads, meaning extra fuel energy and greenhouse gas emissions; and as a consequence leads an increase in unit operation cost.

This paper aims developing a novel approach for heat exchanger fouling monitoring; which is based on heat exchanger simulation, real time process data monitoring and fouling prediction. In this respect a software was being developed and applied in both crude oil distillation unit preheat train exchangers and diesel hydro-processing unit feed/effluent heat exchangers in TÜPRA[a] refinery. In this software, the clean heat transfer conditions of the heat exchangers are simulated by considering the effects of physical and chemical properties such as temperature, pressure, composition, etc. of both tube and shell side flows of each exchanger considering the heat exchanger design specifications. Then the fouling factor, (or fouling resistance, RF) is being monitored from the difference between this clean heat transfer coefficient and the actual heat transfer coefficient obtained from real time process data. The simulation results were found to be consisted with real time operational data taken from the refinery. Going one step further, according to the fouling models -which is developed by relating the unit charge type (from storage tank or cracked, etc.) with the fouling factor models-the maintenance period of the units can be optimized. In this respect, the developed genetic and simulation algorithm run simultaneously in terms of economic aspects. The objective function of the optimization states a comparison between marginal process cost and maintenance cost due to unit shut down and extra fuel cost arising due to fouling; and the optimization constraints are defined according to the operational variables. The optimization algorithm was applied to the diesel hydroprocessing unit and the optimum cleaning time period of the heat exchangers were found.

(220d) Economic, Safety and Environmental Multi-Objective Optimisation of Chemical Pathway Selection

Lam, H. L.,
Ng, D. K. S., University of Nottingham Malaysia
Leong, Y. K.,

Economic, Safety and Environmental Multi-Objective Optimisation of Chemical Pathway Selection

Designing a more sustainable process has become a critical issue with stricter environmental legislation, increasing public concern, repeating plant accidents and escalating environmental problems. Thorough literature review has shown that many researchers have concentrate on implementing either safety or environmental concern into design process. However, only few works have been done on integrating both criteria simultaneously together with economic goal and none is done using multiple indicators. In this paper, a systematic hierarchical framework which focuses on multi-objective optimisation of all three considerations simultaneously using various safety and environmental indexes has been developed. Following the proposed framework, Spreadsheet Assessment Preliminary Analysis (SAPA) is developed to assist process designer in selecting, evaluating and optimising process alternatives available simultaneously during preliminary design stage easily and quickly. Both the framework and SAPA is illustrated by using hydrogen cyanide (HCN) production case study. The developed tool is able to be exteded to different type of process plant.

(220e) Troubleshooting of Sncr Technology for NOx Removal

Feng, M., Southwest Research Institute

Nitric oxides (NO_x), including nitric oxide (NO) and nitrogen dioxide (NO_2), are toxic pollutants from combustion. Most commonly used post-combustion treatment methods include selective catalytic reduction (SCR), selective noncatalytic reduction (SNCR), reburning, and the combination of these technologies. SNCR technology does not need a catalyst, can be used for NO_x emission control. It has the potential of solving the problems inherent with selective catalytic reduction (SCR) technology, e.g., high catalyst cost, high maintenance cost, and sensitive to impurities in flue gas. This paper will address possible problems encountered in the SNCR technology, including narrow reaction temperature zone, short residence time, flow distribution, NH_3 slip, SO_2 effect, water vapor effect, solid particles formation, high temperature materials usage, and corrosion, etc. Computer fluid dynamics simulation can be sued as process optimization tool in the plant design. Application of different reducing agents for NOx removal by SNCR technology, including carbon monoxide, hydrogen, char, hydrocarbons, and oxygenates, etc, will be summarized.

(220b) Development Of LAB-Scale Process For Neopentyl Glycol

Hashmi, A.,

Neopentyl glycol (NPG) is a highly reactive, yet thermally stable, chemical intermediate. NPG is a valuable starting material for the manufacture of polyester resins, polyurethanes, synthetic lubricants, polymers and plasticizers. The key raw materials required for the manufacture of NPG are isobutyraldehyde and formaldehyde, and the most prevalent method for commercial production is aldol condensation of the said aldehydes followed by the hydrogenation of the resulting aldolate. This paper describes the development of a novel lab-scale process technology for Neopentyl Glycol. The cross-aldolization of isobutyraldehyde and formaldehyde was performed using phase transfer catalyst (PTC) as sole catalyst followed by hydrogenation over an environmentally friendly Copper-Zinc oxide catalyst. The PTC promotes reaction between two aldehydes located in different phases of a mixture by transferring one reactant across the interface into the other phase, so that reaction can proceed efficiently. The PTC method is an alternative to commonly used alkali hydroxides, alkali carbonates, ion exchange resins and tertiary amine catalysts. We present the benefits achieved in terms of high productivity, better quality, smaller catalyst charge, control exothermicity, improved safety, enhanced environmental performance, etc. from the industrial viewpoint.

(218b) Scale-Up of Continuous Reactors Using Phenomenological-Based Models

Monsalve-Bravo, G. M., Universidad Nacional de Colombia
Moscoso-Vasquez, H. M., Universidad Nacional de Colombia
Alvarez, H., Universidad Nacional de Colombia

SCALE-UP OF CONTINUOUS REACTORS USING PHENOMENOLOGICAL-BASED MODELS

For years scaling up has been a sort of art in which the expertise, rules of thumb, trial and error, particular solutions and subjective decisions have been used to obtain a proper result at a new scale [1]. Traditionally, chemical processes scale-up is performed based on dimensional analysis perspective, similarity principle criteria, rules of thumb, empirical relations constructed from a set of process data, and in recent years using phenomenological-based models supported by empirical relationships [2]. This shows that it does not exist an infallible method for scaling up a given process, and that a huge effort and proficiency is required to achieve an acceptable outcome of the process at industrial scale [1].

Within the drawbacks of traditional scale-up methods, the main problem is to make a complete list of the relevant independent variables. Here the science meets the art: the choice of the variables is highly subjective and without any rigor [3], [4]. In addition, during scale-up mass transfer, energy transfer, momentum transfer and kinetic reaction could change significantly between scales, and traditional scale-up methods do not take into account how avoiding these changes [5]. This forces the designer to decide by expertise which mechanism is the most important (dominant) of the process. Here, the real governing regime (Dominant Operating Regime, *DOR*) is unknown and there is no direct method for finding it [1], [2]. This fact illustrates that traditional scale-up methods do not always leads into the best commercial unit design.

It is well-known that a model can represent precisely a given system. However, the main inconvenient to use a model for scaling up a process is its validation at several scales; especially because the model parameters vary with the change of scale [6], [7]. As a way to overcome this inconvenient and taking into account that a phenomenological-based model is a fundamental tool to comprehend the behavior of a given system, this work presents a new methodology for scaling up continuous processes using Hankel matrix for capturing process dynamic behavior and scaling up a given process including its dynamic behavior. The methodology herein proposed is based in a previous one developed by Ruiz and Alvarez [2], but overcoming its limitations.

The proposed procedure uses an extension of the discrete form of a control tool called Hankel matrix which has been widely used in model reduction, systems identification, digital filter design [8], and recently in controllers design for establishing inputs and outputs pairings [9]. This methodology uses a Phenomenological-Based Semiphysical Model (*PBSM*) to represent the process system and Hankel matrix for: *(i)* representing the dynamic behavior of the process (design variables - state variables terms), *(ii)* determining the effect of the design variables as a whole over each state variable and *(iii)* scaling up the process maintaining its dynamics hierarchy, establishing the real scale factors.

For determining the dynamics hierarchy the State Impactability Index (*SII*) of each state variable is computed. This index represents the impactability of process designs variables as a whole over a *k-th* given state [9]. According to this, the main dynamic (slowest dynamic) is the state variable with the highest *SII* in the process, i.e., the governing mechanism is intimately related with the main dynamic allowing finding the Dominant Operating Regime (*DOR*). *SII* is computed by means of Singular Value Decomposition (SVD) of the matrix obtained by multiplying the discrete observability (O_b) and controllability (C_o) matrices; such product is known as the Hankel matrix.

The methodology consists of the following steps:

1. Define capacity variable at the current and new scales.

2. Obtain a *PBSM* of the process under investigation.

3. Define the state variables, design variables, synthesis parameters and design-variables-dependent parameters from the model obtained in second step.

4. Fix the Operating Point (*OP*), considering that it must be stable.

5. Explicit each design-variables-dependent parameter as a function of the design variables.

6. Linearize the obtained model around the *OP*, considering the design variables as manipulated inputs.

7. Numerically condition *B* and *C* matrices obtained from linearization to make inputs (design variables) and outputs (state variables) dimensionless and normalized.

8. Discretize the linear model obtained.

9. Compute the Observability, Controllability and Hankel matrices from the discrete model.

10. Compute singular values from Hankel matrix.

11. Compute *SII* of each state variable.

12. Repeat ninth to twelfth steps until the capacity variable's value is equal to its value at the new scale.

13. Compare *SII* values at each scale. If *SII* values at the current scale (*cs*) are approximately equal to *SII* values at the new scale (*ns*), continue with the next step. On the opposite case, a successful scale-up is not possible from the process synthesis established.

14. Compute design-variables-dependent parameters as a function of the design variables at the *ns* (scale factors).

15. Simulate the process with the synthesis parameters, design variables and design-variables-dependent parameters values at the *ns* in order to prove if scale-up task is successfully developed.

The proposed methodology is applied to a non-isothermal polymerization reactor for scaling up this process from $0,1m^3$ to $1,0m^3$. To do this, *SII* index is computed considering two cases: *(i)* the overall heat transfer coefficient required (U_r) and *(ii)* the overall heat transfer coefficient available (U_a) (maintaining the geometrical similarity). In Table 1 the *SII* values of the state variables is shown at the *cs* and *ns*. Here, the dynamics hierarchy in both cases is the same, i.e., *SII* values of $x_3>x_4>x_1>x_5>x_2>x_6$ respectively.

Table 1: *SII* values at $0,1m^3$ and $1,0m^3$, with U_r and U_a.

State Variables	SII_{cs}	SII_{ns} **with** U_r	SII_{ns} **with** U_a
x_1	*6,52*	*6,52*	*13,57*
x_2	*1,61*	*1,61*	*2,83*
x_3	*125,34*	*125,34*	*261,97*
x_4	*72,47*	*72,47*	*151,65*
x_5	*2,74*	*2,74*	*5,85*
x_6	*1,08*	*1,08*	*1,13*

In both cases, x_1 (which represents total inactive polymer concentration) is the main (slowest) dynamic. In addition, in Table 1 is shown that the *SII* values remain constant with U_r and increase with U_a. If it is considered that changes in *SII* values show the deterioration of the *OP* with changes in the operating scale, in Table 2 state variables values at *OP* and the polymer average molecular weight (*y*) are compared for both cases at each scale demonstrating this deterioration.

Table 2: *OP* comparison at $0,1m^3$ and $1,0m^3$ with U_r and U_a.

Variables	**Values at *cs***	**Values at *ns* with** U_r	**Values at *ns* with** U_a	**SI Units**
x_1	*1,3Ã—*10^{-1}	*1,3Ã—*10^{-1}	*5,0Ã—*10^{-4}	$kmol/m^3$
x_2	*5,5*	*5,5*	*1,3*	$kmol/m^3$
x_3	*2,0Ã—*10^{-3}	*2,0Ã—*10^{-3}	*5,2Ã—*10^{-1}	$kmol/m^3$
x_4	*4,9Ã—*10^{1}	*4,9Ã—*10^{1}	*4,7Ã—*10^{2}	kg/m^3
x_5	*335*	*335*	*430*	*K*

x_6	*297*	*297*	*301*	*K*
y	*25000*	*25000*	*907*	*kg/kmol*

This shows that preserving geometrical similarity do not guarantee the same yield of the process at the new scale, and that in order to satisfy the energy requirements of the process, the reactor's jacket must be redesigned by adding baffles. It is also showed that by maintaining the dynamics hierarchy it is possible to obtain the same process yield accomplished at *cs*.

The main contribution of this work is the integration of an index (*SII*) to the scale-up task which allows the establishment of the real scale factors of a given process, and the *DOR* (governing mechanism) which was quite difficult to establish with traditional methods. In addition, the proposed procedure allows scaling up the process maintaining the same dynamics hierarchy through changes of scale.

REFERENCES

[1] A. P. J. Sweere, K. C. A. M. Luyben, and N. W. F. Kossen, ?Regime Analysis and Scale-Down: Tools to Investigate the Performance of Bioreactors,? *Enzyme and Microbial Technology*, vol. 9, no. 7, pp. 386?398, Jul. 1987.

[2] A. A. Ruiz and H. Alvarez, ?Scale-up of Chemical and Biochemical Processes Using a Phenomenological Model (in Spanish),? *InformaciÃ³n TecnolÃ³gica*, vol. 22, no. 6, pp. 33?52, 2011.

[3] M. C. Ruzicka, ?On Dimensionless Numbers,? *Chemical Engineering Research and Design*, vol. 86, no. 8, pp. 835?868, Aug. 2008.

[4] M. RÃ¼disÃ¼li, T. J. Schildhauer, S. M. A. Biollaz, and J. R. van Ommen, ?Scale-Up of Bubbling Fluidized Bed Reactors - A Review,? *Powder Technology*, vol. 217, no. 0, pp. 21?38, Feb. 2012.

[5] M. Ehly, P. J. Gemperline, A. Nordon, D. Littlejohn, J. K. Basford, and M. De Cecco, ?Scale-Up of Batch Kinetic Models,? *Analytica Chimica Acta*, vol. 595, no. 1?2, pp. 80?88, Jul. 2007.

[6] A. Bisio and R. L. Kabel, *Scaleup of Chemical Processes Conversion from Laboratory Scale Tests to Successful Commercial Size Design*. New York: John Wiley & Sons, Inc., 1985, p. 699.

[7] V. J. Inglezakis and S. G. Poulopoulos, ?6 - Reactors Scale-up,? in *Adsorption, Ion Exchange and Catalysis Design of Operations and Environmental Applications*, Amsterdam: Elsevier, 2006, pp. 523?550.

[8] H. Yin, Z. Zhu, and F. Ding, ?Model Order Determination Using the Hankel Matrix of Impulse Responses,? *Applied Mathematics Letters*, vol. 24, no. 5, pp. 797?802, May 2011.

[9] L. A. Alvarez and J. J. Espinosa, ?Methodology Based on SVD for Control Structure Design (in Spanish),? *XIII Congreso Latinoamericano de Control AutomÃ¡tico*, p. 6, 2008.

(220f) Evaluation of Mixed-Absorbent in Chemical Absorption-Biological Reduction Integrated Approach for NOx Removal From Flue Gas

Liu, N., Insititute of Industrial Ecology and Environment
Zhang, L., Zhejiang University
Li, W., Insititute of Industrial Ecology and Environment

Evaluation of mixed-absorbent in chemical absorption-biological reduction integrated approach for NO_x removal from flue gas

*Nan Liu, Lei Zhang and Wei Li**

Institute of Industrial Ecology & Environment, Zhejiang University,

Hangzhou, 310027, China

Abstract:

Nitrogen oxides (NO_x) are usually released from the combustion of fossil fuels, which has contributed to various environmental problems, such as global warming [1]. So far, several technologies have been developed for NO_xremoval. These include selective catalytic reduction (SCR) and selective non-catalytic reduction (SNCR) [2, 3]. However, these conventional control systems suffer from high cost and the risk of secondary pollution [4].

Presently, a new method, called chemical absorption-biological reduction integrated method for NO_x removal from flue gas, is a promising approach [5]. It involves NO enhanced absorption by ferrous complex (Fe(II)L, L: EDTA or Citrate *et. al.*) and reduction of ferric complex (Fe(III)L, caused by oxygen in flue gas), Fe(II)L-NO by bacteria. Previous studies showed that the Fe(II)EDTA system had obvious superiority, but it is still restricted by some factors, such as high market price and poor environmental-friendly characteristics.

A review of research indicated that Citrate was originally used in the desulfurization of flue gas, such as SO_2 absorption by sodium citrate [6] and H_2S removal by Ferric Citrate Complex [7]. Our preliminary researches [8] also confirmed that Fe(II)Cit (Cit: Citrate) was cheaper to absorb NO, and could also be used as carbon source to lessen cost, but the capability of NO removal is much lower than Fe(II)EDTA.

In present studies, an innovation mixed-absorbent of Fe(II)Cit/Fe(II)EDTA (Fe(II)L) system was introduced to make up for each other's deficiencies. After studying the mixed- absorbent system through long-term operation, some influencial factors, such as the concentration of NO, O_2, gas flow rate, were also investigated in the biofilter.

The study was conducted under anaerobic conditions with a pH value of 6.3-7.0. The bioreactor [9] was supplied with NO (250ppm), O_2 (3%, v/v), CO_2 (15%, v/v), N_2 (simulated actual flue gas). The molar ratio of Fe(II)Cit to Fe(II)EDTA in the mixed absorbent was 3:1 [10]. The initial concentration of Fe(II)L was 10mmol·L^{-1}. Bacterial strains used in this bioreactor were *Pseudomonas* sp. DN-2 (GenBank accession No.: DQ811956) [11] and *Escherichia coli* FR-2 (GenBank accession No.: EU693909)[12] with terminal electron acceptors of Fe(II)L-NO and Fe(III)L respectively, which were isolated and enriched in our previous study [11,12].

After a long time operation period, all the recycled liquid solution was replaced by the fresh Fe(II)L (10mmol·L^{-1}) medium. Then the effects of key factors on the biofilter performance were investigated. The biofilter was loaded with simulated flue gas. The influence of NO, O_2, gas flue rate was determined by a gradual increase of the concentration. R_E was introduced to stand for NO removal efficiency.

The Fe(II)L concentrations were determined by modified 1, 10-phenanthroline colorimetric method at 510nm [12]. The inlet and outlet NO concentrations were measured with a chemiluminescent NO_x analyzer (Thermo, Model 42*i*-HL). Microscopic analyses were made by an environment scanning electron microscope (Philips, model XL30E, SEM). All data shown in this paper were the mean values of duplicate or triplicate experiments.

The result shows that R_E maintained at about 90% during 170h operation, while Fe(II)L concentration kept at about 6mmol·L^{-1}. To investigate the growth of biofilm on the polyethylene fillers, samples of polyhedral sphere were taken out from the upper and bottom sections of the medium bed for analysis by SEM at the end of long-term operation stages. The results show that the microorganisms can grow well on filter medium in comparison with the contrast one. The formation of the biofilm was enhanced with the biofilter operation. Subsequently, high level of $R_E$could be attributed to the abundance of microorganisms in the biofilm.

The kinetic behavior accords with the Michaelis-Menten model. It can be concluded that Fe(II)L existed as absorbent definitely improved the NO diffusing process from gas phase to liquid phase.

The effect of NO concentration on R_E was determined in the range of 250-800ppm during 28h operation. It was observed that $R_E$maintained over 90%, which had no significant changes in current experimental conditions. The experimental result indicates that the system has potential to remove NO at high loading conditions.

Flue gas from power plants usually contains amount of oxygen depending on combustion conditions. Oxygen has undesirable effects on NO absorption with the iron chelate. The concentration of NO was 250ppm and O_2 was increased from 1% to 5% (v/v) in 34h periodic steps. It can be observed that the system was highly effective (most up to 90%) for R_E. The R_E can return to steady level within 6h during each step in O_2 inlet concentration. The Fe(II)L concentration maintained over 5mmol·L^{-1}. As the O_2 concentration increasing from 1% to 5% (v/v), Fe(II)L reduced by microorganisms substantially kept a balance with Fe (II) oxidation with O_2. So it appeared that the oxidation of Fe(II)L is not strongly affected by the fluctuation of oxygen concentration. The system shows a well adaptability to the shock load of the oxygen.

The experiment was conducted to evaluate the interferences of gas flow rate in the process. R_E and Fe(II)L concentrations remained steady, even if the inlet gas flow rate was increased (1-2.5L·min^{-1}).

In conclusion, a mixed-absorbent constituted by Fe(II)Cit and Fe(II)EDTA was utilized to achieve the CABR process. The system could maintain at 90% of NO_x removal efficiency. In addition, the biofilter presented a high degree of flexibility in the changes of operating conditions (inlet concentration fluctuations of NO, O_2, and gas flow rate). The mixed-absorbent provides the possibility for continuous removal of NO from flue gas in industrial application prospect.

References

[1] A. Richter, Increase in tropospheric nitrogen dioxide over China observed from space. Nature 2005, 437, 129-132.

[2] X. Gao, Y. Jiang, Y. Zhong, Z.Y. Luo, K.F. Cen, The activity and characterization of CeO_2-TiO_2 catalysts prepared by the Sol-gel method for selective catalytic reduction of NO with NH_3. J. Hazard. Mater. 2010, 174 (1-3), 734-739.

[3] J. Jirat, F. Stepanek, M. Marek, M. Kubicek, Comparison of design and operation strategies for temperature control during selective catalytic reduction of NO_x. Chem. Eng. Technol. 2001, 24 (1), 35-40.

[4] M. Radojevic, Reduction of nitrogen oxides in flue gases. Environ. Pollut. 1998, 102 (1), 685- 689.

[5] L. Gao, X.H. Mi, Y. Zhou, W. Li, A pilot study on the regeneration of ferrous chelate complex in NO_xscrubber solution by a biofilm electrode reactor. Bioresour. Technol. 2011, 102 (3), 2605-2609.

[6] F. James, B. Sune, Citrate solution absorbs SO_2. Chem. Eng. 1980, 87 (12), 88-89.

[7] A.J. Bard, Encyclopedia of electrochemistry of the elements ( ). New York: Marcei Dekker Inc. 1976, 307.

[8] W. Li, N. Liu, L. L. Cai, J. L. Jiang, J. M. Chen, Reduction of Fe(III) chelated with citrate in an NO_x scrubber solution by *Enterococcus* sp. FR-3. Bioresource Technol. 2011, 102, 3049-3054

[9] B. H. Lu, Y. Jiang, L. L. Cai, N. Liu, S. H. Zhang, W. Li, Enhanced biological removal of NOx from flue gas in a biofilter by Fe(II)Cit/Fe(II)EDTA absorption. Bioresource Technol. 2011, 102 (17), 7707-7712.

[10] N. Liu, B. H. Lu, S. H. Zhang, J. L. Jiang, L. L. Cai, W. Li, Y. He, Evaluation of Nitric Oxide Removal from Simulated Flue Gas by Fe(II)EDTA/Fe(II)citrate Mixed Absorbents. Energ. Fuel 2012, 26 (8), 4910-4916.

[11] S.H. Zhang, W. Li, C.Z. Wu, H. Chen, Y. Shi, Reduction of Fe(II)EDTA-NO by a newly isolated *Pseudomonas* sp. strain DN-2 in NO_x scrubber solution. Appl. Microbiol. Biotechnol. 2007, 76 (5), 1181-1187.

[12] W. Li, C.Z. Wu, S.H. Zhang, K. Shao, Y. Shi, Evaluation of microbial reduction of Fe(III)EDTA in a chemical absorption-biological reduction integrated NO_x removal system. Environ. Sci. Technol. 2007, 41 (2), 639-644.

(218c) BPA Process Development Using Aspen Modeling: The "Interface" Between Technology & Operations

Youssef, A. A., SABIC
Cosby, T. J., Tennessee Technological University
Khadilkar, M., SABIC Innovative Plastics

An ASPENTECH™ simulation workbook was built for SABIC™ at its BPA Plant in Mt. Vernon, IN. The workbook was designed to allow users to troubleshoot, optimize, and improve the BPA Process. The excel interface provided a powerful tool for the manufacturing and operations teams to run cases testing the process capability and potential changes from the baseline operating conditions. Examples include the impact of reaction feed compositions and catalyst activity on the plant production as well as the effect of the purge treatment in the BPA recovery section. The workbook served to illustrate the effects of key process input changes.

The excel worksheet is connected to the overall BPA ASPEN PLUS™ process model. There is an individual worksheet for the reaction, dehydration, separation, and recovery sections. However, each section is run on the same model so changes in one section affect the results in another. Highlights of the interface and its utilizations are emphasized in this presentation.

SABIC is a trademark of Saudi Basic Industries Corporation or its affiliates

ASPENTECH and ASPEN PLUS is a trademark of ASPEN TECHNOLOGY, INC.

(220s) Study On Preparation of Conjugated Linoleic Acids With Alkali Catalyst From Natural Unsaturated Fatty Acid Methyl Esters

Ma, L., Jiangnan University
Fan, Y., Jiangnan University
Fang, Y., Jiangnan University

Conjugated linoleic acids (CLAs) are a group of positional and geometrical isomers of conjugated octadecadienoic acids, which are naturally unsaturated fatty acid with conjugated double bonds. As is shown to inhibit carcinogenesis and atherosclerosis, participate in fat metabolism and enhance immune function, CLAs have been widely applied in food, pharmaceutical, health care, cosmetics and many other fields. Chemical modification methods, therefore, to prepare CLAs from natural unsaturated fatty acids have attracted more and more attention nowadays, which mainly include dehydration of ricinoleic acid in castor oil, alkali- or metal-catalyzed isomerization of linoleic acid, enzyme-catalyzed transformation and microorganism fermentation. In this paper, breaking traditional synthetic methods, it studied the synthesis of CLAs with methyl linoleate as starting material. The CLAs after synthesis were characterized by GC and UV-Vis. The content of CLAs reached 90 % with sodium hydroxide as catalyst, which concentration was 20 % compared with methyl linoleate, at 170 ^{o}C for 4 h.

(220g) How Detailed Mass Balances in Biomass Biorefining Processes Can Help in the Development of An Efficient Sugar-Platform Biorefinery

Arantes, V., University of British Columbia
Olsen, C., University of British Columbia
Saddler, J. N., The University of British Columbia

A sugar-platform Biorefinery is a facility that integrates fractionation, separation and conversion processes to derive soluble, monomeric sugars from carbohydrate-containing feedstocks such as forest biomass and agricultural residues. The sugar streams are then used as the major raw material for production of a wide range of bioproducts such as biochemicals and transportation liquid fuels. Different technologies and process configurations with varying degrees of maturity and complexity have been proposed to generate sugar intermediate streams. However, biomass pretreatment followed by enzymatic hydrolysis form the key upstream process of a number of pilot, demonstration, and planned commercial biomass-to-bioproducts industries aiming to access and release sugar residues from biomass feedstocks.

While, it is becoming increasingly apparent that sugar recovery yields, chemical composition of the sugar streams as well as chemical/biochemical inputs (e.g. enzyme dosage) all critically affect the techno-economic viability of a sugar-platform Biorefinery operation, determining the optimal upstream process conditions that maximize sugar recovery yield while minimizing biocatalyst input and the formation of byproducts that affect sugar downstream processing has proved challenging. This is primarily because of the mutual interdependence of the pretreatment and enzymatic hydrolysis steps as operating conditions that allow for high sugar recovery yields during pretreatment leads to low sugar yield during enzymatic hydrolysis at low biocatalyst input. Therefore, the upstream pretreatment and enzymatic hydrolysis process steps are essentially a compromise.

Because of the heterogeneous nature of biomass feedstocks and the multitude and complexicity of the biocatalyst mixtures make kinetic studies a difficult task and with limited practical application, it is anticipated that empirical process modeling, despite overlooking parameters related to reaction kinetics and transport phenomena, can provide practical insight for research and development as well as a useful tool for process control and optimization. This presentation will first discuss the accuracy of a series of empirical modeling approaches [1], some derived from pulping processes, to predict sugar recovery yields and the chemical composition of the sugar streams after biomass pretreatment. Modeling the dependence of sugar yields on biocatalyst dosage during enzymatic hydrolysis will be also discussed. In particular, examples will be presented on how detailed mass balances obtained from empirical modelling of the pretreatment and enzymatic hydrolysis steps could help to also assess the impact that industrially relevant process factors might have on the overall sugar yields and biocatalyst input.

[1] Olsen C, Arantes V, Saddler JN. 2012. The use of predictive models to optimize sugar recovery obtained after the steam pre-treatment of softwoods. Biofuels, Bioproducts and Biorefining 6 (5):534-548.

(220h) Application Of Natural Products As Dispersants/Flocculation Of Asphaltenes

Aguiar, J. I. D. S. D., Institute of Macromolecules
Moreno, M. M., EQ/UFRJ
Mansur, C. E., Federal University of Rio de Janeiro

Asphaltenes are the oil components that can form aggregates and deposit, causing financial damage to the oil industry during processing. To solve this problem, stabilizers can be used to inhibit the deposition of asphaltenes. On the other hand, the oil industry may get benefits of using flocculants of asphaltenes in oil processed to increase the value of oil or to increase compatibility between different oils. In addition, nowadays there is an increasing concern about the sustainability of industrial processes.

The objective of this study is to evaluate the efficiency of more than ten natural products extracted from fruit (e. g. clove oil and sweet orange oil) as flocculants/dispersants of asphaltenes and determine a natural product to replace a commercial additive that isn't renewable and is often more harmful to people and the environment that the natural products. For this purpose, asphaltenes proceeding from a distillation residue was extracted and characterized by elemental analysis (to determine their elemental composition), by microcalorimetry (to determine their range of solubility parameter) and by NMR (to measure its content of aromatic carbons). Precipitation tests were conducted using natural products, analyses were made with a spectrophotometer in the visible ultraviolet region at a wavelength of 850 nm with the optical path of 2 mm and 5 mm. Natural products have shown promising performance for the use as dispersants / flocculants of asphaltenes.

(220t) Taking Into Account Transport Parameter Evolutions in Composting Process to Improve Energy Consumption

Debenest, G., Institut de Mecanique des Fluides de Toulouse
Druilhe, C.,

Composting process refers to a biological treatment of wastes thanks to the oxygen conveyed by the flow, humidity within the sludge and temperature increase due to biological activity.

There is a strong coupling existing between transport processes and biological reactions and this generates also a compaction process of the solid porous matrix.

This talk will be dedicated to the understanding of the compaction effects on the whole process. This compaction comes from the initial rearrangement (in the first hours) of the porous medium. Due to biodegradation reactions and drying, some pieces of the porous matrix are decomposed in smaller parts and a second slow compaction process appears (from the second day to the end of the aeration process). It is then necessary to characterize those effects during these two periods.

A coupled experimental and numerical macroscale approach (Darcy's scale) has been followed. We have obtained a data bank for effective transport parameters but also mass balances. Then, we have make use of those relations using our numerical tool and compare the numerical results to the experimental ones.

This shows a good agreement and allows us to propose new developments in order to improve the efficiency of the industrial process. Some perspectives of aeration limitations will be given in order to reduce the consumption cost of energy.

(220i) A New Sulfuric Acid Recycle Routine in TiO2 Pigment Production

Liang, B., Sichuan University

A new sulfuric acid recycle routine in TiO_2 pigment production

Liang Bin, Li Chun

Sichuan Uinversity, Chengdu 610065, P. R. China

In China, more than 99% of the titanium resources exist in the form of ilmenite located in Panzhihua area, Sichuan province. Nowadays about 60% of the world's TiO_2 pigments are manufactured by the chlorination route, in which natural or synthetic rutiles, or titanium-rich slags are used as feeding stocks. The shortage of natural rutiles has encouraged the researches to convert ilmenite into synthetic rutile. Therefore, China produces about 1.2 million tons of TiO_2 pigment per year in sulphate process, in which ilmenite is firstly leached with concentrated sulphuric acid and then the Ti-contained solution is purified and hydrolyzed. The TiO_2 is obtained by calcining the hydrolysate.

The sulphate process brings with about 8 tons of `20 wt.% spent sulphuric acid for each ton of TiO_2 production (Liang et al., 2005; Li et al., 2007d), which results in severe environmental problems. In fact, the process can be simply described as: concentrated acid ® dissolution of ilmenite ® purification ® hydrolysis ® dilute acid. Recycling dilute acid can solved the emmision problem.

In a commercial process, the ilmenite was leached with a concentrated sulfuric acid of 98% at about 200°C. Our leaching kinetic experiments were carried out with an initial acid/ilmenite (wt) ratio of 500:15, a sulfuric acid concentration of 15.4M and temperatures from 100° to 198°C. Based on the leaching kinetics model obtained, simulation results showed that the optimal initial acid concentration was 85%. It indicated that the 20% waste acid can be usd as the digestion acid by mixing the 98% concentrate acid. The practical operation reduced about 15% emission of waste acid.

In view of recycling the waste acid, we need to further decrease the digestion acid concentration. Using a low acid concentration, however, the digestion rate is too low to practice. Milling ilmenite can activate the ore and therefore it can enhance the digestion of ilmenite in dilute acid. The dissolution of mechanically activated Panzhihua ilmenite at 100°C with a 50% sulfuric acid solution was investigated. With the mechanically activated ilmenite inside a ball mill, leaching conversion at 100°C reached 82.1%. The leaching activation energy decreased from 48.1KJ/mol to 28.4kJ/mol. (Li et al., 2008; Li and Liang, 2008) Furthermore, we investigated the simultaneously milling digestion and we found that a high leaching conversion of ~80% can be reached under an ore/acid ratio closed to the commercial operation. The initial acid concentration of 50% can reduce more than 60% of waste acid emission.

Based on Shrinking model, the activity energies are:

Raw ore	**55.7kJ/mol.**
Milling 40min	**31.4kJ/mol**
Milling 60min	**29.4kJ/mol**
Milling 120min	**23.7kJ/mol**

WelhamF

33.1kJ/mol®39.1kJ/mol

The 20% H_2SO_4 was used to directly dissolve the ilmenite, it can completely avoid the emission of waste acid. We dissolved the mechanically activated Panzhihua ilmenite in 5~20% sulfuric acid solutions and found that both titanium and iron almost dissolved at a stoichiometric ratio. However, the final dissolving degree of Ti was very low due to the hydrolysis of the dissolved titanium during the digestion process. The digestion results bring us a chance to leaching ilmenite with dilute acid and extract the iron species, while the titanium hydrolyzed to form a solid product. The residue was detected with XRD and the result showed that the product was rutile TiO_2. It means that the digestion of ilmenite with extremely dilute acid can not produce the titanium solution suitable for the hydrolysis, but it brings with a new method to produce synthetic rutile which can be used as the feedstock of the chlorination TiO_2 pigment production.

MillingF 2h TemperatureF 100°C

A

B

C

D

E

Waste sulfuric acid is a severe environmental challenge of sulphate routine TiO_2 pigment production. The solving method is reducing the initial digestion acid concentration and recycle the hydrolysis spend acid. The mechanically activation of ilmenite greatly enhances the leaching process and makes it possible to leach the ilmenite with dilute acid. It will be a sustainable process of TiO_2 pigment production.

Reference

[1] Chun Li, Bin Liang, Hao Song, Xiao-qing Wang. Preparation of porous rutile titania from ilmenite by mechanical activation and subsequent sulfuric acid leaching. Microporous & mesoporous materials. 2008, 115 : 293–300.

[2] Chun Li, Bin Liang. Study on the mechanochemical oxidation of ilmenite. Journal of Alloys and Compounds. 2008, 459(1-2) 354-361.

[3] Chun Li, Bin Liang, HaiYu Wang.Preparation of synthetic rutile by hydrochloric acid leaching of mechanically activated Panzhihua ilmenite, Hydrometallurgy, 91(1-4): 121-129(2008)

[4] Chun Li, Bin Liang. Dissolution of mechanically activated Panzhihua ilmenite in dilute solutions of sulphuric acid. Hydrometallurgy. 2007. 89:1-10.

[5] Chun Li,Bin Liang,Zibin Wu, Linghong Guo. Effect of mechanical activation on the dissolution of Panzhihua ilmenite. Minerals Engineering. 2006.19:1430|1438.

[6] Chun Li, Bin Liang, Sheng-Pin Chen. Combined milling-dissolution of Panzhihua ilmenite in sulfuric acid. Hydrometallurgy. 2006. 82: 93-99.

[7] Bin Liang, Chun Li, Chenggang Zhang, Yongkui Zhang. Leaching Kinetics of Panzhihua Ilmenite in Sulfuric Acid.Hydrometallurgy.2005.76(3-4): 173-179.

(220j) Algal FAME Production With A NOVEL Surfactant Based Catalyst In A Reactive Extraction

Salam, K. A., Newcastle University

Harvey, A. P., Newcastle University

Velasquez-Orta, S.,

ALGAL FAME PRODUCTION WITH A NOVEL SURFACTANT BASED CATALYST IN A REACTIVE EXTRACTION

*Kamoru A. Salam, **Sharon B. Velasquez-Orta, ***Adam P. Harvey

*Kamoru.salam@ncl.ac.uk, **Sharon.velasquez-orta@ncl.ac.uk, ***adam.harvey@ncl.ac.uk

School of Chemical Engineering and Advanced Materials (CEAM), Newcastle University,

NE1 7RU, United Kingdom.

ABSTRACT

One major challenge in conventional algal biodiesel production is the high energy requirement of the drying and solvent extraction steps. The high solvent requirement, (usually hexane) for the oil extraction in this process also has a significant environmental impact. Alternatively, biodiesel production could take place via *in situ* transesterification (?reactive extraction?). In this process fatty acid methyl ester (FAME) is produced directly from micro-algal biomass using methanol containing a catalyst. This approach is simpler and potentially more cost-effective, because of its elimination of the solvent extraction. The current study involved development of a novel surfactant catalyst (zirconium dodecyl sulphate) for usage in the ?reactive extraction? of FAME from *Chlorella vulgaris* and *Nannochloropsis occulata*. Additionally, the FAME yield produced when using sodium dodecyl sulphate and H_2SO_4 catalyst was investigated. Microalgae was also characterised before and after the reactive extraction in terms of carbohydrate and protein content. For *Chlorella* strains a maximum 67wt% FAME yield was reached at 24 hours by the sodium dodecyl sulphate and H_2SO_4 catalyst; 54wt% FAME yield by the H_2SO_4 catalyst. For the *Nannochloropsis* a maximum 75wt% FAME yield was reached at 12 hours by the sodium dodecyl sulphate and H_2SO_4 catalyst; 44wt% FAME yield by the H_2SO_4 catalyst. At 12 hours the FAME yields produced by the zirconium dodecyl sulphate were 67wt% for *Nannochloropsis* and 8wt% for the *Chlorella*. In conclusion this finding shows that:

- The inclusion of sodium dodecyl sulphate in H_2SO_4 significantly improved the FAME yields obtained from *Nannochloropsis* and *Chlorella*.
- A zirconium dodecyl sulphate catalyst performed better with *Nannochloropsis* than in *Chlorella*.
- The micro-algae maintains almost all the protein content and some of the

Carbohydrate after the *in situ* transesterification

- The reaction time and yields were significant function of algae strain and catalyst.

Key words: reactive extraction, microalgae, catalyst, FAME, H_2SO_4, surfactant

INTRODUCTION

During plant photosynthesis, bio-oils are produced as triglycerides which have a high combustion heat.. One way of utilising this energy is trans esterification. The use of refined oil for trans esterification is however not economical as ~ 88% of total production cost of this conventional two steps biodiesel production is ascribed to refined oil feedstock (Haas, 2006). Similarly biodiesel production from non-food oil crops is not sustainable. Microalgae are one of the sustainable biofuel feed stocks because of their compelling advantages. They are non-food biofuel feed stocks. They can be cultivated on non-arable land. Algae can be grown on the sea or wastewater. They can also be used for capture highly concentrated CO_2 .

Production of FAME from micro-algae via *in situ* trans esterification is simple and potentially more economical since ~90% of the process energy is accounted for by the solvent extraction and drying steps in the conventional trans esterification (Lardon *et al.*, 2009). Previous studies have demonstrated the feasibility of obtaining greater FAME conversion from such *in situ* trans esterification than from a conventional two-step approach (Lewis *et al.*, 2000; Vicente *et al.*, 2009. The major drawback of this method is the requirement of a large amount of methanol. This is necessary since methanol plays a dual role: it acts as an oil extractor and as a reactant. Moreover, the need to recover methanol from the products streams may add additional cost to the process.

This present research may render the process more economical by producing co-associating value products such as carbohydrate, protein and other components in the micro-algal residue after *in situ* trans esterification.

MATERIALS AND METHOD

Procedure for *in situ* trans esterification

This is a modification of Velasquez-Orta *et al.* (2012). The following process conditions were maintained in the *in situ* trans esterification: 100 milligram of microalgae, temperature: 60°C, mixing rate: 450 rpm, mole ratio of catalyst to oil: 8.5:1, mole ratio of methanol to oil: 600:1 (Either the surfactant catalyst, Sodium dodecyl sulphate and H_2SO_4 or H_2SO_4 catalyst) and methanol was added to each tube consecutively and it was transferred to a shaking incubator (IKA KS 4000i control). The *in situ* trans esterification was run for time ranged from 30minutes - 36hours. After each completed reaction the tube was kept in a freezer for at least 6 hours to stop the reaction. The algae residues were separated from the bulk liquid by centrifugation. The bulk liquid mixture of methanol, FAME and by-products was transferred in pre-weighed tubes. The final weight of the bulk liquid was recorded for each tube and the FAME concentration was measured by gas chromatography.

RESULTS AND DISCUSSION

FAME yield profiles obtained in *an in* situ trans esterification of freeze dried cells of *Nannochloropsis occulata* is shown figure 1. It was observed that the FAME profiles for the three catalysts used general increased with reaction time. This is the general trend of a normal reaction. However, a drop in the FAME yield after 12 hours was observed with the SDS and H_2SO_4 catalyst. This could be due to conversion of FAME to free fatty acid by a side reaction. It was also observed that ~70% increase in FAME yield was produced by SDS and H_2SO_4 more than that of H_2SO_4 at 12 hour. FAME yield produced by the Zirconium dodecyl sulphate catalyst was also more than that of H_2SO_4 catalyst at 24 hours.

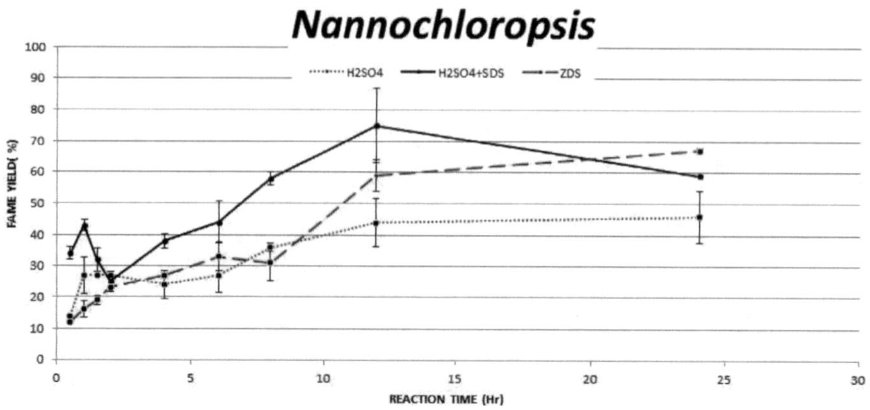

Reactive Extracted FAME Yield Profiles of the *Nannochloropsis*
Molar ratio of methanol to oil = 600; molar ratio of catalyst to oil = 8.4; Agitation rate = 450 rpm; Temperature = 60°C; SDS: Sodium dodecyl sulphate; ZDS: Zirconium dodecyl sulphate

Fig.1: **Reactive Extracted FAME Yield Profiles of the *Nannochloropsis***

Molar ratio of methanol to oil = 600; molar ratio of catalyst to oil = 8.4; Agitation rate = 450 rpm; Temperature = 60°C; SDS: Sodium dodecyl sulphate; ZDS: Zirconium dodecyl sulphate

FAME yield profiles obtained in an *in situ* trans esterification of spray dried cells of *Chlorella vulgaris* is shown figure 2. It was observed that the FAME profiles for the three catalysts used general increased with reaction time. However, after 4 hours the FAME yield produced by the ZDS decreased with reaction time. Similarly, a relatively low performance of the ZDS catalyst was observed in this strain. This could be due to the difference in the cell wall Chemistry between the two strains. The inclusion of SDS with H_2SO_4 enhanced the FAME yield obtained from this species. At 24 hours ~24% increase in FAME yield was obtained by Sodium dodecyl sulphate and H_2SO_4 than that of H_2SO_4 catalyst.

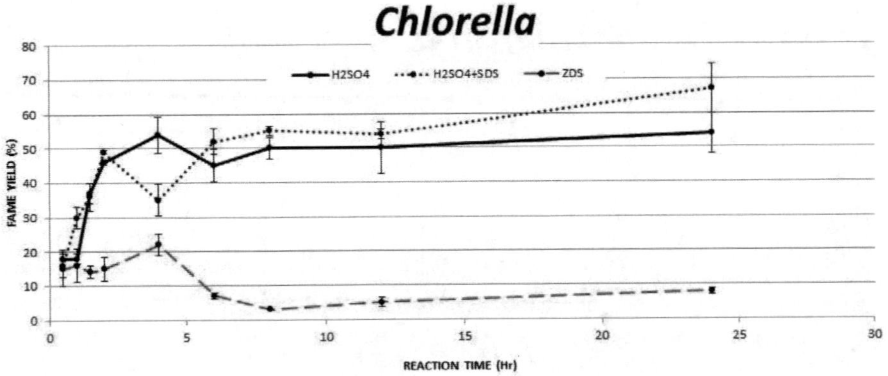

Reactive Extracted FAME Yield Profiles of the *Chlorella*
Molar ratio of methanol to oil = 600; molar ratio of catalyst to oil = 8.4; Agitation rate = 450 rpm; Temperature = 60°C; SDS: Sodium dodecyl sulphate; ZDS: Zirconium dodecyl sulphate

Fig2. : **Reactive Extracted FAME Yield Profiles of the *Chlorella vulgaris***

Molar ratio of methanol to oil = 600; molar ratio of catalyst to oil = 8.4; Agitation rate = 450 rpm; Temperature = 60°C; SDS: Sodium dodecyl sulphate; ZDS: Zirconium dodecyl sulphate

The protein and carbohydrate content of the *Nannochloropsis* residue was measured after the reactive extraction. The results of the quantification are shown in Figure 3 and 4 respectively. It can be seen from the figure 3 that the biomass maintains almost all the protein content after the reactive extraction. The residual carbohydrate in contrast was lower than their original values before the reactive extraction. This shows that some of the carbohydrates have been hydrolysed during the reaction to simple sugars and other co-associated products.

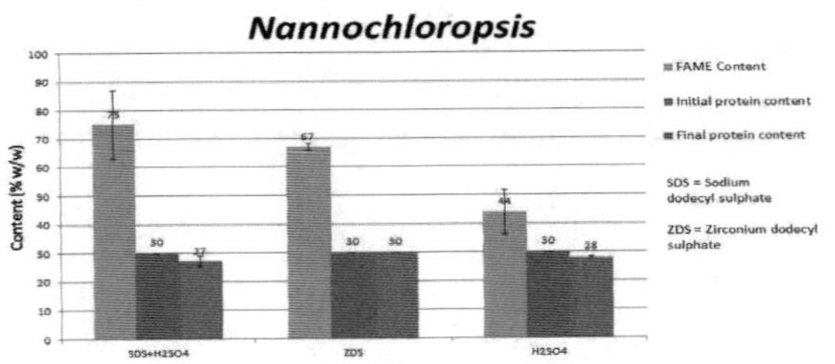

Protein content of the *Nannochloropsis* before and after the reactive extraction

Fig.3: Protein content of the *Nannochloropsis* before and after the reactive extraction

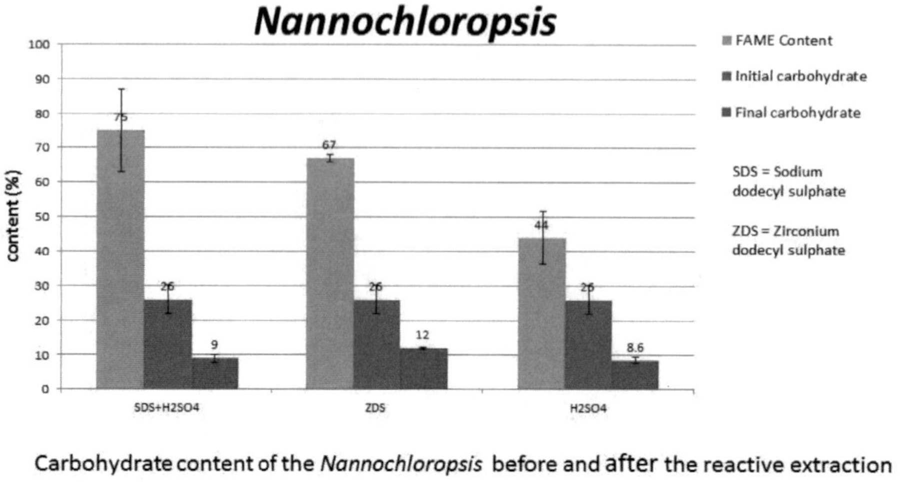

\s

Fig. 4: Carbohydrate content of the *Nannochloropsis* before and after the reactive extraction

The protein and carbohydrate content of the *Chlorella* residue was also measured after the reactive extraction. The results of the quantification are shown in Figure 5 and 6 respectively. It can be seen from the figure 5 that the biomass maintains almost all the protein content after the reactive extraction. The residual carbohydrate however was lower than their original values before the reactive extraction. This also shows that some of the carbohydrate has been hydrolysed during the reaction to simple sugars and other co associated products.

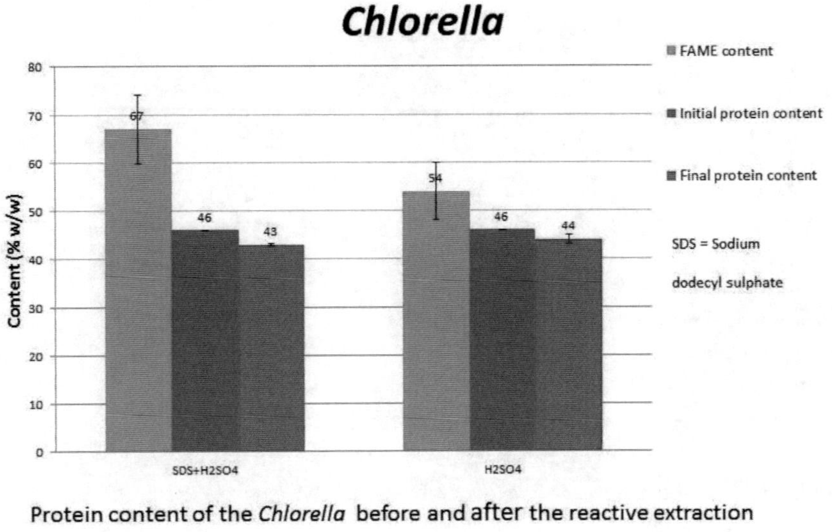

Fig.5: Protein content of the *Chlorella* before and after the reactive extraction

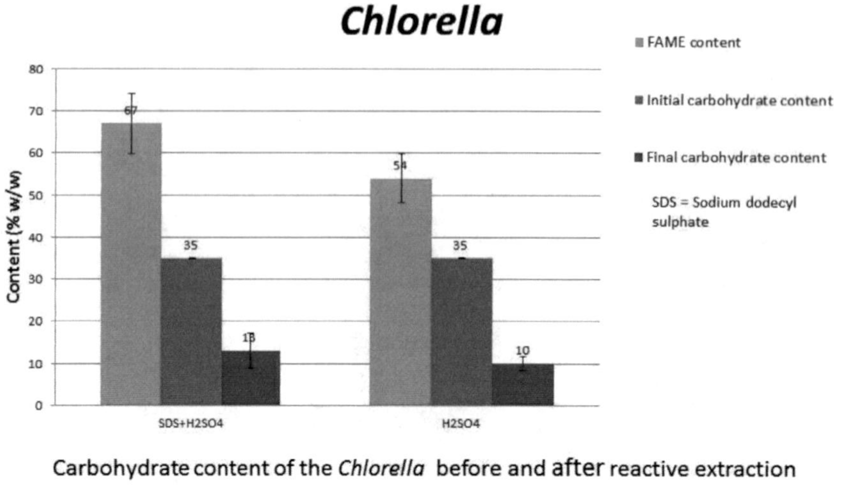

Fig.6: Carbohydrate content of the *Chlorella* before and after reactive extraction

Table 1: Initial Biochemical Content of the microalgae strains

Component (wt %)	*Chlorella*	*Nannochloropsis*
Carbohydrate[a]	35	26
Protein[b]	46	30

[a]Measured using acid digest by Gerhardt *et al.* [1]

[b]Measured using Nitrogen protein conversion factor for microalgae by Lourenc *et al.* [2]

REFERENCES

[1] Gerhardt P. , Murray R. G. E., Wood W. A., Krieg, N. R. (1994)." Methods for General and Molecular Bacteriology." ASM ,Washington D. C , p518.

[2] Sergio o. Lourenc, Elisabete Barbarino, Paris L. Lavixn, Ursula M. Lanfer Marquez and Elizabeth Aidar (2004) ?Distribution of intracellular nitrogen in marine microalgae: Calculation of new nitrogen-to-protein conversion factors? Eur. J. Phycol., 39 (1): 17 ? 32.

[3] Velasquez-Orta, S. B., Lee, J. G. M. and Harvey, A. (2012). "Alkaline *in situ* transesterification of Chlorella vulgaris." *Fuel*, 94, 544-550.

[4] Haas, M. J., McAloon, A. J., Yee, W. C. and Foglia, T. A. (2006). ?A process model to estimate biodiesel production costs.? *Bioresource Technology*, 97, 671-678.

[5] Lardon L., Sialve B., Steyer J. and Bernard O. (2009). ?Life-Cycle Assessment of Biodiesel Production from Microalgae.? *Environmental Science and Technology*, 17, 6475-6481.

(220k) System Study On Mass Production of Biodiesel From Microalgae With Nutrients Recycling Process

Kakimura, M., Tokyo University of Agriculture and Technology
Fushimi, C., Tokyo University of Agriculture and Technology
Kato, D.,
Iwasaki, H.,
Mitsumori, T.,
Yamada, K., Center for Low Carbon Society Strategy, Japan Science and Technology Agency (JST), The University of Tokyo

Mass production of biofuel from microalgae is a promising technology because of their high productivity of bio-oil and rapid growth rate compared with other biomass plants. However, a large amount of nutrients such as N, P and K is required for large-scale culturing of microalgae. So far, few studies have proposed a large-scale bio-diesel production system considering cost of nutrients or their recycling.

In this study, at first we calculated cost of nutrients for mass production of microalgae with and without their recycling. It was found that the bio-diesel costs at least 1400 yen/L-bio-diesel (1 US$=100 yen) without nutrients recycling even if some of the nutrients are supplied by waste water. This clearly suggests that most of the nutrients must be recycled from microalgae residue after bio-oil extraction.

Based on this result, we have proposed four processes (Process 1: bio-oil extraction by Hydrothermal Liquefaction (HTL) and bio-diesel production by supercritical (s.c.) methanol method; Process 2: bio-oil extraction by HTL and bio-diesel production by alkali method; Process 3: bio-oil extraction by hexane and bio-diesel production by s.c. methanol method; and Process 4: bio-oil extraction by hexane and bio-diesel production by alkali method) with nutrients recovery after bio-oil extraction. The material balance, energy balance and total costs of the four processes were calculated. The results show that the Process 1 is most feasible because the energy consumption of bio-diesel production was lowest (0.19 MJ/MJ-biodiesel) among the four processes. The capital cost of the Process 1 was calculated to be 1.3×10^2 yen/L-bio-diesel when the annual expenditure rate for the fixed cost is assumed to be 0.2 of the capital cost excluding the land cost. The variable cost was calculated to be 1.1×10^2 yen/L-bio-diesel.

(220m) Streamlining the Harvest of Algae and Conversion of Algal Lipid

Lin, C. C., University of Utah
Hong, A., University of Utah

Streamlining the Harvest of Algae and Conversion of Algal Lipid

Ching-Chieh Lin, P.K. Andy Hong

Department of Civil and Environmental Engineering, University of Utah, Salt Lake City, UT, USA

Abstract

Algae-derived biodiesel is seen with great potential to be sustainable and able to contribute to global demand for transportation fuels. Currently, harvesting and dewatering microalgae are a significant bottleneck in developing algal biodiesel. No single best method has been identified for harvesting and dewatering microalgae effectively and economically, which generally requires steps such as concentration, filtration, dewatering, grinding, and solvent extraction. In this study, combing well practiced sand filtration and ozonation processes, we have collected and ruptured microalgae into lipid in the liquid phase, combining the need to dry and extract separately, thus enabling the immediate conversion step into biodieseal. Microalgae, *Chlorococcum aquaticum*, was acidified to pH 3.3 to promote agglomeration prior to biomass collection by sand filtration. After their retention in the sand bed, addition of methanol and ozonation for 2 min completed cell rupture for release of cell contents. The lipid in methanol solution was collected and converted in biodiesel. The results indicated successful collection of agglomerated microalgae via sand filtration (95%), effective rupturing of microalgae via ozonation (87%), and a 16% lipid yield from the microalgae, as well as restoration of filtration velocity of the sand bed via prolonged ozonation. GCMS identification of the extract showed primary products in the forms of long-chain largely saturated hydrocarbons of 16 to 20 carbons, which were converted into biodiesel. The integral process streamlines individual steps in the procurement of algal lipid from suspension that improves over existing energy-intensive methods.

Key words used: microalgae, biodiesel, harvest, dewater, sand filtration

(220l) Recovering a Liquid-Lignin Phase From Paper Mill Black Liquors

Velez, J., Clemson University
Thies, M. C., Clemson University
Klett, A. S., Clemson University

Recovering a Liquid-Lignin Phase from Paper Mill Black Liquors

Julian Velez and Mark C. Thies*

Department of Chemical and Biomolecular Engineering, Center for Advanced Engineering Fibers and Films, Clemson University, Clemson, SC, 29634-0909 USA.

* *Corresponding author. E-mail: mcths@clemson.edu*

In collaboration with Liquid Lignin Company, we have developed a process for recovering from paper mill black liquors a liquid-lignin phase of reduced ash content in the form of a low-viscosity, easy-to-process fluid. The properties of this liquid lignin are such that a very low ash (< 1.0 wt %) lignin can readily be produced from it.

In this work, we investigated both hardwood and softwood black liquors under a wide range of conditions, including temperature, pressure, and pH, in order to determine the range over which this unique liquid-lignin phase forms. Measurements of several fundamental lignin properties, including molecular weight, hydroxyl content, melting point, and sodium content, were carried out in order to relate precipitation conditions to characteristics of the lignins produced.

A semi-batch operating mode was used, with CO_2 being bubbled into a black-liquor charge in order to bring about the desired phase split. Experiments were carried out in a stirred, 2-L Parr reactor setup especially modified for our work. The CO_2 was fed to lower the pH of the black liquor from 13.5 to 9.5. For each set of experimental conditions, a heavy, liquid-lignin phase and a light, spent black liquor phase were generated, with samples of each phase being collected for further analysis.

Implications of the results obtained to date for the production of very low ash lignins for applications such as biofuels and value-added bioproducts will also be discussed.

(220u) New Applications of FTIR Spectroscopy for Process Control and Process Development

Sellin, D.,

New Applications of FTIR Spectroscopy for Process Control and Process Development

D. Sellin[1], M. Bothe[2], J. Fagaschewski[1], S. Baum[1], G. K. Padmanabhan[1], G. M. Shibabaw[1], S. Bohne[3], L. Hilterhaus[1], H. K. Trieu[3], M. Schlüter[2], A. Liese[1]

[1]Institute of Technical Biocatalysis, [2]Institute of Multiphase Flows, [3]Institute of Microsystem Technology

Abstract

Infrared spectroscopy is a powerful method for qualitative and quantitative determination of substances. Infrared radiation causes molecular vibrations, the resulting absorption spectra give information on molecular structure and concentration.

In the mid infrared range (4000 400 cm ^{1}), absorption bands are more intense and less broad in contrast to near-infrared (12000 – 4000 cm^{-1}), therefore an assignment of characteristic functional groups to single bands is possible. The information content is higher, making multivariate regression methods such as Partial Least Squares (PLS) regression more powerful[1].

In process analytics, the optical setup of Attenuated Total Reflectance (ATR) in combination with PLS chemometrics allows quantitative inline determination of all reactants simultaneously. Moreover, even high viscous multiphase systems on industrial scale can be monitored with a single probe in presence of up to 4 phases, as shown by the biocatalytic synthesis of myristyl myristate in a bubble column[2].

In process development, local determination of characteristic key parameters is possible, e.g. mass transport coefficients of CO_2in bubble columns or residence time distributions by means of IR-active tracers.* Here, first results from measurements with multiple ATR-probes will be presented. The local determination of parameters adds a layer of space resolution to the measurements, creating new possibilities for the development and validation of scale independent models.

In case of transmission measurements, due to the relatively high absorbance of samples in the mid-infrared, a layer thickness of a few micrometers is required, hampering sample preparation. The limited applicability was overcome by the development of IR-transparent microreactors and microtiter plates, which are presented here. These devices open new possibilities for the screening of enzymatically catalyzed reactions.

*This project is supported by the Bundesministerium für Bildung und Forschung within the scope of the interdisciplinary project "Chemical Processing: Multiscale Modeling of Multiphase Reactors (Multi-Phase)", which arose from a pilot project out of the network „Campus Blasensäulen“.

[1] Kara, S., Müller, J. & Liese, A. (2011). Online Analysis Methods for Monitoring of Bioprocesses. *Chemistry Today* **29**(2) 38-41.

[2] Müller, J. J., Neumann, M., Scholl, P., Hilterhaus, L., Eckstein, M., Thum, O. *et al.* (2010). Online Monitoring of Biotransformations in High Viscous Multiphase Systems by Means of FT-IR and Chemometrics. *Anal. Chem.* **82**, 6008–6014.

(220n) A Generalized QSPR Model for Henry's Law Constant of Hydrogen in Organic Compounds

Odafin, C. E., Oklahoma State University
Gebreyohannes, S. B., Oklahoma State University
Yerramsetty, K. M., Oklahoma State University
Neely, B. J., Oklahoma State University
Gasem, K. A. M., Oklahoma State University

Henry's Law constant is a key physical property, which determines the distribution of gases in liquids. Hydrogen gas has a wide range of applications in chemical processes. For instance, in biomass conversion to biofuels, hydrogenation reactions are applied in the reduction of the oxygen content of the biofuel. Hydrogen solubilities in various solvents have been documented widely, and empirical models for Henry's law constants have been determined; however, these data are expensive to obtain and the empirical models are mostly suited for the available data. Hence, the need exists for generalized models for predicting Henry's constants. These generalized models will be applicable for *a priori* predictions of compounds lacking experimental data or for newly and yet to be synthesized compounds.

In this work, the objective was to develop QSPR generalized models for the *a priori* prediction of Henry's constant in various organic solvents. Specifically, the model results have been applied to hydrogenation reactions in a biphasic catalytic reactor for the upgrading of bio-oil to biofuels.

Hydrogen solubility and Henry's constant data were collected for a diverse range of compounds at various temperatures and pressures. The generalized equation of state (GEOS) software was used in the regression and evaluation of the available data. Quantitative structure-property relationship (QSPR) methodology was then applied to generalize the model parameters of the regressions. A representative database was employed in the model development and validation. The goal is to obtain hydrogen solubility predictions within 2-3 times the average experimental errors.

(220a) Detoxification of Dilute-Acid and Ionic-Liquid Pretreated Wheat Straw Hydrolysates to Improve Succinic Acid Fermentation

Shao, H., University of Toledo
Varanasi, S., The University of Toledo
Relue, P., The University of Toledo
Viamajala, S., University of Toledo

Biomass hydrolysate may be used for the production of succinic acid via fermentation. However, concentrated hydrolysate produced from dilute acid or ionic liquid pretreatment contain several inhibitors, especially low molecule weight phenolic compounds (LMWPCs) that strongly inhibit the growth of *A. succinogenes*. In this study, we evaluated solid phase adsorption (activated charcoal, ion-exchange resin, and polymeric resin) and a novel liquid-liquid extraction (LLE) process for detoxifying wheat straw hydrolysate. The results showed that although solid phase adsorption removed the major inhibitors, a 3%-16% sugar loss reduced the overall succinic acid yield (62% to 69%, g-succinic acid/g-sugar). The LLE process using tri-n-octylphosphine oxide (TOPO) or tri-n-octylamine (TOA) efficiently removed nearly 100% of LMWPCs and more than 60% of the total phenolics from both hydrolysates with no sugar loss. Succinic acid fermentation yield on the LLE-detoxified hydrolysates was nearly 75% (g-succinic acid/g-sugar), which was as good as the yield from the model sugar control.

(220o) Energetic and Environmental Assessment of Biomass to Ethanol Processing in the Sugar-Cane Industry

Kahwaji Janho, M., Cleveland State University
Shuster, Q. A., Cleveland State University
Gatica, J. E., Cleveland State University
Hernandez, M. R., Universidad Nacional de Tucuman
Mele, F. D., Universidad Nacional de Tucuman
Colombo, M., Universidad Nacional de Tucuman

Bio-fuels have come to be recognized as possible, environmentally friendly alternatives to fossil fuels. This work studies a Bio-ethanol production process complementing a sugar mill, through computer simulations, using ASPEN PLUS® as the simulation engine.

The process flowsheet examined includes three major processing stages typical of distillery installations for fuel-grade production in typical sugar-cane processing facilities: (1) the fermentation stage, (2) the separation stage, and (3) the dehydration stage.

Using plant data from a bio-ethanol pilot plant in Northwestern Argentina (NOA) a kinetic model is formulated for a fed-batch molasses fermentation process. The model for this unit, along with a user-defined separation module, is then integrated into a commercial process simulator. Different schemes meeting equivalent performance standards in terms of purity and recovery are examined in this paper.

Particular attention is given to energetic integration by performing a pinch-analysis to each of the fundamental stages to find a global optimum. Optimization is achieved by minimizing a given objective function that accounts for bio-ethanol recovery, separation efficiency, and minimization of effluents. Life Cycle Analysis is used to complete the assessment of operating parameters/conditions on environmental effects.

(220p) Mechanism of Resistance of a Yeast to Sulfur Dioxide and the Genetic Improvements for Its Application in Advanced Solid-State Fermentation for Fuel Ethanol Production Using Sweet Sorghum

Zhang, L., Tsinghua University
Li, S., Tsinghua.University
Fu, X.,
Qi, L.,
Zu, X.,

Compared to sugar cane, sweet sorghum has great potentials to be a principal energy crop worldwide due to its supreme agronomics advantages, such as fast growing, broad cultivation regions, and strong tolerance to drought, waterlogging, salinity, and alkalinity. Recently, sweet sorghum has been used as the feedstock for a pilot-scale fuel ethanol production using continuous advanced solid-state fermentation (ASSF) developed by Tsinghua University, China. This first demonstrated the high feasibility of applying ASSF at large scale for fuel ethanol production. For further industrialization and commercialization of ASSF, the sustaining of sugars in sweet sorghum stalks is a prerequisite for year-round production of fuel ethanol using ASSF. Accordingly, extensive studies have focused on the long-term storage strategies of sweet sorghum. However, the high inhibitory activities of antiseptics to yeast fermentation and poor homogenicity between the solid sweet sorghum and the antiseptics result in no effective storage methods so far to support the commercialization of sweet sorghum ethanol using ASSF. Most recently, we developed a storage strategy using sulfur dioxide (SO_2) as the antiseptics that can keep sugars in sweet sorghum stalk with no more than 5% of sugar loss over 6 months. Meanwhile, we obtained a *S. cerevisiae* strain THS2 that shows a strong tolerance to SO_2. After a serial of SO_2 adaption cultures, TSH2 is capable of growing in the SO2-containing media up to 1600 ppm, which shows much higher tolerance of yeast to SO_2 than that of ever reported. In this study, we investigated the mechanism of TSH2's tolerance to SO_2, and applied the strategies of genetic and evolutionary engineering to reduce the lag phase of TSH2 in SO_2-containing media. We can expect that this strategy is capable of supporting the commercialization of sweet sorghum ethanol production using ASSF.

(220r) Extra-Heavy OIL Fractions Upgrading Characteristics In A DUAL Circulating Fluidized BED Reactor

Seo, M. W., Korea Institute of Energy Research
Guahk, Y. T.,
Ra, H. W., Korea Institute of Energy Research
Yoon, S. J., Korea Institute of Energy Research
Kim, Y. K., Korea Institute of Energy Research
Lee, J. G., Korea Institute of Energy Research
Nho, N. S., Korea Institute of Energy Research
Kim, S. D., Korea Advanced Institute of Science and Technology

Among 2 trillion barrels of whole oil reservoir, the portion of unconventional oils (heavy oil, extra-heavy oil, oil sand bitumen, and asphalt) reaches 53.3 % of oil reservoir. However, due to constraints such as high density, viscosity, metal, and sulfur contents, the production and supply of heavy oil are quite limited. The upgrading process is essential for the use of unconventional oils. The extra-heavy oil fractions upgrading process, which consists of a rapid thermal pyroylzer (RTP) of extra-heavy oil and gasifier of RTP residue to produce syngas as well as supply heat to the pyrolyzer are developed in KIER (Korea Institute of Energy Research. Unreacted carbon in RTP residue from the pyrolyzer will be used as a feedstock to the gasifier. The silica sand acts as a bed material circulating from the pyrolyzer and RTP residue gasifier.

In this study, the dual circulating fluidized bed reactor which consists of a riser (0.078 m-I.D. X 7 m-high) and a bubbling fluidized bed (0.2 m-I.D. X 2.1 m-high) having 1 BPD production capacity was designed and constructed for extra-heavy oil upgrading process. The bitumen and VR (Vacuum Residue) were used as a feedstock, whereas, the air and steam were used as gasifying agents. The pyrolysis and gasification performances such as oil production rate, product gas composition, carbon conversion, gas yield, heating value of product gas are determined.

(218a) Assessment of an Eulerian-Eulerian Multiphase CFD Framework for a Three Dimensional Modeling of a Trickle Bed Reactor in Counter-Current Operation

Moreno, S.,
Matos, E. M., UNICAMP
Souza, V.,
Ferreira, A. S.,
Monteiro, C. A. A.,
Nunhez, J. R., UNICAMP

Commercial HDT processes usually operate in a trickle-bed regime, with co-current downward flow of gas and liquid over a randomly fixed bed of catalyst particles where reactions take place. It is well known that sulfur removal is strongly inhibited by the competitive adsorption effect of H2S at the sulfided active sites of the catalyst. As a consequence, it is important in these process to maintain the concentration of H2S as low as possible during the reaction to achieve a low sulfur content product at the reactor outlet. According to Ancheyta et al. (2007), a more convenient profile of H2S concentration can be provided by operating the reactor in counter-current mode, for instance, introducing the feed at the top and H2 at the bottom of the reactor. Thus, in the lower section of the reactor, the H2S concentration is lower and H2 is greater, promoting higher reaction rates. The main problems on the counter-current operation are a lower efficiency on the gas-liquid contact; a difficulty in preventing flooding conditions and a higher pressure drop. In order to investigate the above processes in more detail, a three-dimensional fluid-dynamic study of a counter-current trickle bed reactor was carried out using CFD (Computational Fluid Dynamics) techniques with emphasis on determining the pressure drop and the distribution of liquid hold-up in the bed. A CFD tridimensional model based on an Eulerian – Eulerian multiphase approach was used to model the hydrodynamics of the pseudo two-phase flow in a trickle bed reactor (TBR) in counter-current operation. The closure terms for the phase interactions have been based on the fluid-fluid interfacial force balance concept (Attou et al. 1999), which was used and validated to simulate co-current HDT reactors by Gunjal e Ranade (2007). Radial variation of porosity was estimated using the Klerk (2003) distribution. The above set of model equations were implemented inside the commercial software ANSYS-CFX Release 13. The model provided information about phase distribution, velocity profiles, pressure drop liquid hold up for a isothermal trickle bed reactor at laboratory scale in counter-current operation.

(220q) Direct Solar-Powered Biomass Gasification Using Low-Temperature Steam

Manenti, F., Politecnico di Milano

Ravaghi-Ardebili, Z., Politecnico di Milano

Direct Solar-Powered Biomass Gasification Using Low-Temperature Steam

Flavio Manenti, Zohreh Ravaghi-Ardebili*

Politecnico di Milano

Dipartimento di Chimica, Materiali e Ingegneria Chimica ?Giulio Natta?

Piazza Leonardo da Vinci 32

20133 Milano, Italy

**Corresponding author. Phone: +39 (0)2 2399 3287; Fax: +39 (0)2 7063 8173; Email: flavio.manenti@polimi.it*

ABSTRACT

This work is aimed at coupling the concentrating solar technology with the biomass gasification. The main issue is that the biomass gasification occurs at temperatures of the order of 800-900°C, whereas the available technology in concentrating solar power plants using thermal fluids allows to achieve temperatures of 550-600°C. Different elements are needed to be effective: (1) the detailed characterization of biomass pyrolysis and the successive gas phase reactions; (2) the selection of appealing biomass gasification technologies (i.e., traveling grate, countercurrent); (3) the integration with discontinuous energy sources (concentrating solar); (4) the definition of proper thermal energy storage; and (5) the study of effective operational procedures to achieve direct solar powered biomass gasification.

KEY-WORDS

Biomass gasification; concentrating solar; biomass pyrolysis; energy integration; thermal energy storage

INTRODUCTION

Biomass is one of the most promising feedstock able to satisfy the increasing demand for renewable energy, biofuels, and green chemicals (Cucek et al. 2010, Kleme? et al. 2010, Lam et al. 2010a, Lam et al. 2010b, Vaccari et al. 2005). Unfortunately, biomass conversion to bioproducts is tough to be industrially scaled-up due to the complexity of chemical and transport phenomena as well as to be integrated with other renewable sources that can support for energy and steam generation the gasification process. Thus, two critical steps are:

1. the development of mechanistic models capable of describing transport phenomena and reaction kinetics for a better understanding of biomass pyrolysis;

2. the integration of these models at the process scale to develop novel process solutions.

For the former one, detailed chemical mechanisms are needed both for biomass pyrolysis and for the successive gas phase reactions, since they are still unavailable even for major products released such as levoglucosan, hydroxymethylfurfural, and phenolic species (Ranzi et al. 2013). Chemical mechanisms need to be integrated into particle model accounting for transport phenomena, which are critical in predicting global reactor performance. Developing these models is challenging because of the biomass complexity as well as the multi-phase and multi-scale nature of the conversion process (Mettler et al. 2012). For the latter step, the use of two-tanks direct thermal energy storage in concentrating solar plant is an established and appealing technology (Vitte et al. 2012, Manenti and Ravaghi-Ardebili 2013).

A brief overview of these two main topics is given hereinafter.

BIOMASS GASIFICATION

Combustion, gasification, and biomass-to-liquid pyrolysis are some of the main thermo-chemical conversion routes, which can convert an abundant and well distributed feedstock into energy, syngas, bio-oil, and chemicals. One of the main problems when studying this type of feedstock is the intrinsic variability of the biomass composition. As a consequence, it is necessary to properly characterize the biomass, preferably on the basis of few lumped components, which are typical for all the possible feedstock.

The kinetic model adopted is explained in detailed elsewhere (Ranzi et al. 2013) and is based on a multi-step devolatilization and decomposition of the three key-biomass reference species: cellulose, hemicelluloses and lignin. One of the main features of this model is its ability to provide detailed information on yields composition of gas, tar, and solid residue. The kinetic model also involves the char gasification and combustion reactions, with steam and/or air or oxygen, as well as the secondary homogeneous gas phase reactions of the released gas and tar species. The multistep kinetic model has been validated on the basis of thermo-gravimetric data of fine particles, with negligible resistances. Next, the model has been extended to the reactor scale with the analysis of a countercurrent biomass gasifier.

ARCHIMEDE PLANT

Archimede concentrating solar plant uses linear parabolic streams and direct two-tanks thermal energy storage where the molten salt thermal fluid flows inside (Vitte et al. 2012). Energy storage is a major process design and control issue in concentrated solar plants (Manenti and Ravaghi-Ardebili 2013). The intrinsic discontinuous nature of solar energy forces to install units able to store energy under favourable conditions, then to release energy during night or unfavourable conditions. Accordingly, smoothened operations and continuous energy production can be guaranteed. Several technologies are today available for storage; tanks of molten salts, steam accumulators, high thermal capacity solids are among the most important.

SOLAR GASIFICATION

According to the brief descriptions above, the idea is to supply the biomass gasification using the steam generated by the concentrating solar plant. To do so, it would be necessary to produce steam at 900°C, whereas the existing molten-salt solar technology allows to achieve about 600°C. Nevertheless, it is possible to select specific configuration of gasifier that may suit the achievement of the solar-powered biomass gasification. Actually, the use of countercurrent gasifier allows to exploit the high temperature of ashes slowly moving to the bottom of the unit to increase the temperature of the steam entering from the bottom (Figure 1). By doing so, the work is aimed at demonstrating that the lower limit for the inlet steam temperature to keep the ?hot? condition of the biomass gasifier is not higher than the steam temperature obtained from the concentrating solar plant, making it feasible. Oxygen supply is needed to sustain the endothermic reaction of gasification by means of in situ combustion of a portion of biomass. Specific operational procedure is needed to operate the integrated plant. Also, the study will analyse the possibility of cogeneration before supplying the biomass gasifier to further intensify the solution energetically (Sikos and Kleme? 2010, Zhu et al. 2000). The existing GasDS tool for biomass characterization and gasification (Ranzi et al. 2013) is adopted for the dynamic simulation of the countercurrent gasifier and it is integrated in DynSim? suite for process dynamic simulation by Simulation Science, Invensys.

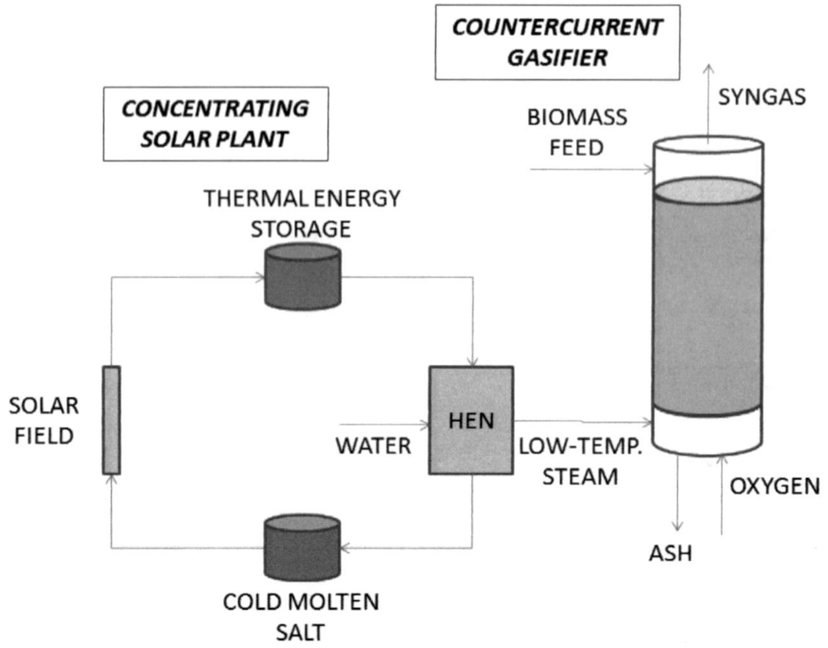

Figure 1. Qualitative layout.

REFERENCES

Cucek, L., Lam, H. L., Kleme?, J. J., Varbanov, P. S. and Kravanja, Z. (2010) 'Synthesis of regional networks for the supply of energy and bioproducts', Clean Technologies and Environmental Policy, 12(6), 635-645.

Kleme?, J. J., Varbanov, P. S., Pierucci, S. and Huisingh, D. (2010) 'Minimising emissions and energy wastage by improved industrial processes and integration of renewable energy', Journal of Cleaner Production, 18(9), 843-847.

Lam, H. L., Varbanov, P. and Kleme?, J. (2010a) 'Minimising carbon footprint of regional biomass supply chains', Resources, Conservation and Recycling, 54(5), 303-309.

Lam, H. L., Varbanov, P. S. and Kleme?, J. J. (2010b) 'Optimisation of regional energy supply chains utilising renewables: P-graph approach', Computers and Chemical Engineering, 34(5), 782-792.

Manenti, F. and Ravaghi-Ardebili, Z. (2013) 'Dynamic simulation of concentrating solar power plant and two-tanks direct thermal energy storage', Energy, (doi: 10.1016/j.energy.2013.02.001).

Mettler, M. S., Vlachos, D. G. and Dauenhauer, P. J. (2012) 'Top ten fundamental challenges of biomass pyrolysis for biofuels', Energy Environ. Sci., 5, 7797-7809.

Ranzi, E., Corbetta, M., Manenti, F. and Pierucci, S. (2013) 'Kinetic Modeling of the Thermal Degradation and Combustion of Biomass', submitted to Chemical Engineering Science.

Sikos, L. and Kleme?, J. (2010) 'Reliability, availability and maintenance optimisation of heat exchanger networks', Applied Thermal Engineering, 30(1), 63-69.

Vaccari, G., Tamburini, E., Sgualdino, G., Urbaniec, K. and Kleme?, J. (2005) 'Overview of the environmental problems in beet sugar processing: Possible solutions', Journal of Cleaner Production, 13(5), 499-507.

Vitte, P., Manenti, F., Pierucci, S., Joulia, X. and Buzzi-Ferraris, G. (2012) 'Dynamic Simulation of Concentrating Solar Plants', Chemical Engineering Transactions, 29, 235-240.

Zhu, X. X., Zanfir, M. and Kleme?, J. (2000) 'Heat Transfer Enhancement for Heat Exchanger Network Retrofit', Heat Transfer Engineering, 21(2), 7-18.

ADDIN EN.REFLIST

(219g) L-Glutamic Acid Crystallization in a Mesoscale Oscillatory Baffled Crystallizer

Abernethy, R. A., Newcastle University
Phan, A. N., Newcastle University
Harvey, A. P., Newcastle University

Background

It is well-known that mixing conditions play a key role in affecting the product properties and qualities during crystallization processes. Many crystallization processes are however carried out in stirred tanks which are known to provide non-uniform mixing conditions. This can result in a product that requires downstream processing to meet the desired specification. In recent years the advantages of oscillatory baffled reactors (OBRs), such as more uniform mixing and heat transfer conditions than in stirred tank reactors, have been identified as being beneficial for use in crystallization. OBRs are capable of achieving plug flow conditions when used in continuous mode of operation which would in theory allow for a continuous crystallization process that has consistent product properties over time. The quality of crystals formed using oscillatory flow mixing conditions has also been found to be higher than those produced using ?conventional' mixing in a stirred tank due to the formation of fewer inclusions (Ristic, 2007). Mesoscale OBRs are OBRs scaled down to a 5 mm internal diameter, with applications in screening of conditions and low volume, continuous manufacture (Reis et al., 2005). Advantages of using this type of reactor for screening include the ability to handle solids, having a compact reactor design and achieving plug flow conditions at laminar flow rates, all of which are applicable to crystallization screening.

Aims

The aims of this research were to determine the viability of continuous crystallization in mesoscale OBRs by studying the crystallization of l-glutamic acid and to investigate the effect that variables including the oscillation conditions have on the crystal properties, such as size statistics and crystal polymorph. These results were compared with those of published crystallization experiments in larger scale OBRs.

Methodology

A solution of known concentration of l-glutamic acid was continuously pumped through a series of mesoscale OBRs, cooling the solution from 80Â°C to 20Â°C with a residence time of 20 minutes. The amplitude and frequency of oscillation in the OBR were varied between experimental runs. Product samples were collected from the outlet of the final mesoscale OBR and were analysed by light microscopy, Raman spectroscopy and X-ray powder diffraction to determine the crystal properties such as the polymorph, crystal habit, and size characteristics. A series of jacketed stirred tanks experiments with the same cooling profile and solution concentration were then used to provide a comparison between oscillatory and ?conventional' mixing.

Results

It was established that cooling crystallization can be successfully carried out using mesoscale OBRs. The most dramatic finding was the occurrence of previously unreported tetrahedral shaped crystals coexisting with the typical α polymorph shape when samples from the mesoscale OBR were viewed under a light microscope as seen in Figure 1. This occurred in the samples produced using the lower oscillation conditions. In samples from the stirred tank experiments this crystal habit was not detected indicating that its existence may be due to the hydrodynamic conditions that occur in the mesoscale OBR. The tetrahedral crystals were able to grow and/or then subsequently change in to the α polymorph shape. This transformation is shown in Figure 2.

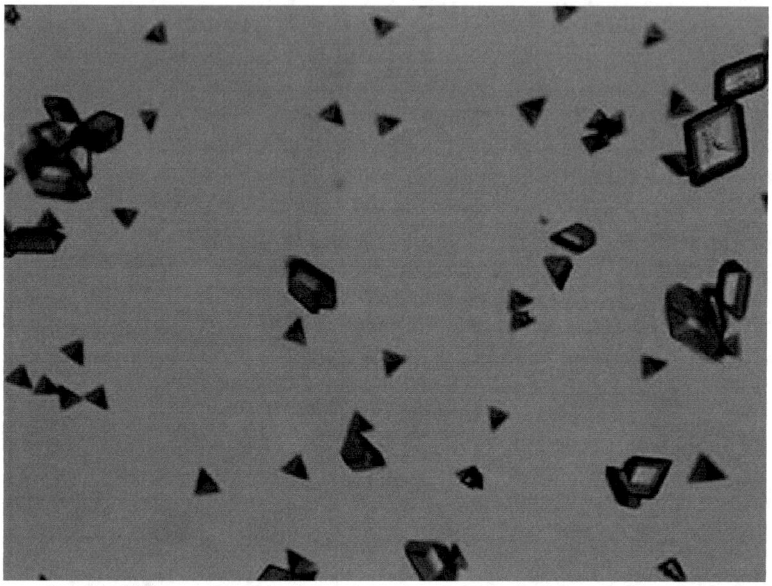

Figure 1 ? Microscope image of crystal products from the mesoscale OBR

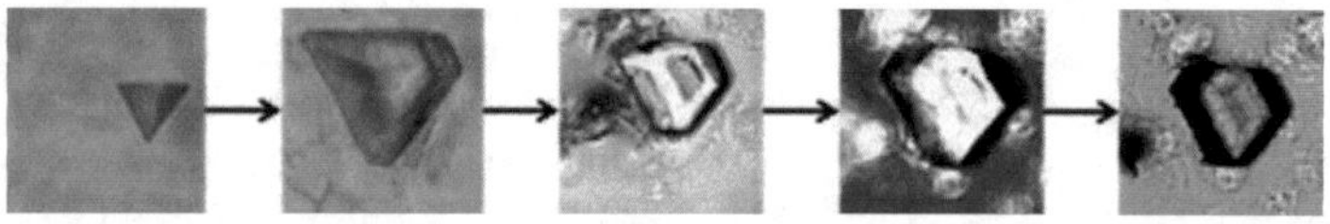

Figure 2 ? Microscope images illustrating tetrahedral crystal transformation (different magnifications are used as images progress)

The crystal size measurements have shown that the l-glutamic acid crystals produced in the mesoscale OBR follow the same trends as in the larger OBRs, with more vigorous mixing conditions producing smaller average crystal sizes. This shows that it should be possible to produce crystals of a given average size by selecting the appropriate oscillation conditions after sufficient experimental data has been obtained. Based on the microscope images, the tetrahedral crystal habit was found to have a critical size of approximately 20 μm at which the transformation into the characteristic α polymorph shape begins to occur.

The samples taken from the stirred tank crystallization experiments showed that many of crystals had some degree of physical damage. This was not observed in the crystals sampled from the mesoscale OBR even at the most vigorous achievable mixing conditions demonstrating that the oscillatory mixing conditions in this situation result less damage to the crystals.

Conclusions

Cooling crystallization has been successfully carried out for the first time in mesoscale OBRs, producing l-glutamic acid crystals. Mean crystal sizes of samples from the mesoscale OBR were found to decrease with an increase in mixing intensity. A previously unreported crystal habit of l-glutamic acid was observed, which was witnessed transforming into the typical α polymorph shape. This tetrahedral crystal habit was not found to occur in experiments performed using a stirred tank suggesting that the mesoscale OBR could be used as a platform to investigate existing crystallization processes and potentially expose novel findings.

References

REIS, N., HARVEY, A. P., MACKLEY, M. R., VICENTE, A. A. & TEIXEIRA, J. A. 2005. Fluid Mechanics and Design Aspects of a Novel Oscillatory Flow Screening Mesoreactor. *Chemical Engineering Research and Design,* 83, 357-371.

RISTIC, R. I. 2007. Oscillatory mixing for crystallization of high crystal perfection pharmaceuticals. *Chemical Engineering Research and Design,* 85, 937-944.

(219a) Process Intensification in Liquid-Liquid Extraction

Aksamija, E., Graz University of Technology
Siebenhofer, M., Graz University of Technology

The design and scale-up of extraction columns (i.e. columns with rotating internals) is still based on empirical rules and practical experience. Implementation of Computational Fluid Dynamics (CFD) into apparatus design is a promising engineering challenge due to the constant increase of calculation power. So far, detailed CFD-simulations of all interconnected phenomena (i.e. turbulence, bubble break-up & coalescence, swarm effects, mass transfer over deformable bubble interfaces, etc.) are still computationally too expansive to allow a reliable quantitative prediction of industrial scale equipment. However, CFD can provide prediction of crucial hydrodynamic parameters (axial dispersion coefficients, hold-ups, drop size distributions, etc.) and their trends in dependency of operation parameters and geometry.

By calibrating the simulation algorithm with experimental data from pilot scale experiments, CFD is used to optimize the internals design by a systematic variation of geometric ratios. For experimental validation the continuous phase flow fields of a *Rotating Disc Contactor* were recorded with *Particle Image Velocimetry* (PIV) and the residence time distribution was obtained by tracer experiments. Based on experimentally validated models and parameters continuous phase simulations combined with the *Euler-Langrange* framework were used to optimize the ratio of the column internals.

With the primary target of minimizing axial dispersion and maximizing intensity/size of the toroidal vortices, geometrical internals ratios were systematically altered in several simulation runs, finally leading to a novel simplified design without stator discs. Subsequent dual phase simulations (Euler-Euler model combined with population balances) were performed to evaluate stability of flow patterns in dual phase operation. Smaller droplets and narrower drop size distributions were observed for the optimized internals geometry, confirming intensification in terms of overall column performance.

Mass transfer experiments were performed to prove operation performance in terms of separation efficiency. Compared with the standard RDC design the height of one transfer unit (HTU) was found to be 40% lower in a pilot scale RDC100 with the simplified internals geometry.

Beside better operation performance the apparatus design without stator internals suggests applicability in complex separation tasks, such as downstream processing of fermentation broths, with explicit tendency to third phase formation in the compartments and the settling zone.

(219d) Three Concepts for Intensification of Nordic Kraft Pulp Processes

Kangas, P., VTT Technical Research Centre of Finland
Kaijaluoto, S., VTT Technical Research Centre of Finland
Maattanen, M.,

Background

The global change of paper consumption is moving the pulp production from northern countries, such as Scandinavia, US and Canada, towards southern hemisphere, to Uruguay, Brazil, Indonesia and China. Thus there is increasing interest for developing competitive future pulp mill concepts applicable in Nordic countries by intensification the current production processes. Proposed concepts of this study are (i) producing sulphur less or lean products, (ii) maximising energy production in pulp mill, and (iii) developing high yield pulps. Typical modern Nordic kraft mill is used as reference for these concepts.

Evaluation of concepts is conducted by literature review, by validating the pulp quality in fibre line with experimental trials (both cooking and bleaching), and by evaluating techno-economic feasibility of concepts by process simulation. In addition, rough investment estimates are given for new concepts.

Sulphur lean concept

The aim of the Sulphur lean concept is to produce birch pulp with low (sulphidity 10 %) or minimal (sulphidity close to 0 %) sulphur content. The main benefit of sulphur lean or free chemical pulping is the possibility to isolate sulphur-free side-products, such as lignin, to be used directly as a fuel or processed further for other uses. Strength is that the pulping process itself is virtually odourless

Pulp quality of sulphur lean concepts is partly better (WRV, bulk, stiffness) and partly worse (wet zero-span tensile strength, fracture toughness and tear strength) than reference pulp. Pulp itself is more difficult to bleach than kraft pulp which leads to higher bleaching chemical consumption and cost. The main disadvantage with concept is the significantly higher demand for causticizing capacity as all hydroxide must "pass through the lime cycle". The attractiveness of proposed concept would be increased further by a breakthrough in auto-causticizing technologies.

Soda black liquors have been known to be challenging to evaporate and the smelt properties are somewhat different than in the kraft process. Black liquor gasification of soda liquors could increase the energy production in chemical pulping or open new possibilities for by-product generation. The gasification of black liquor is evaluated for this sulphur lean black liquor concept.

Energy mill concept

In the energy mill concept a scenario is evaluated where the energy production of the pulp mill is given a greater importance compared to today's situation. As the energy price compared to pulp price is getting higher, there is increased interest in low kappa (<10) pulping. The heating value of carbohydrates (cellulose) is lower compared to lignin, and thus low kappa pulping with yield preserving methods are preferable. AQ and polysulphide are utilized as cooking aids. Cooking to low kappa enables omitting the oxygen stage and possible also chorine stages. (Possible savings in investment and energy costs). Both elemental (ECF) and total (TCF) chlorine free bleaching options were evaluated.

This study also showed that the pulp produced from the Energy mill ECF concept is of reasonable quality. In fact some properties are equally good or even better than the properties of the reference pulp. For example the tensile properties of the ECF pulps were better than those of the reference kraft. Also the very low extractive content in the pulp makes it an interesting pulp for special purposes.

This study also showed that the production costs of the Energy mill TCF concept are too high. Regarding the Energy mill ECF concept the results showed that cooking to kappa 10 and keeping the yield higher than 40 % is possible. If the electricity price would be the same as for electricity produced by wind, the energy mill concept would be economically very close to the reference kraft mill. Including the higher thermal energy output from the Energy mill concept, which in winter months would give increased revenue as well if CHP is installed.

High yield pulp concept

As wood is the major cost component of pulp in Nordic countries, the high yield pulp concepts is aiming to a significant increase in pulp yield. Two-stage pulping concept includes (i) high yield and high kappa cooking for stabilising hemicelluloses in pulp and (ii) novel oxygen delignification aiming to remove the lignin and reduce kappa while obtaining high yield. The proposed concept might include additional processes for recovering dissolved hemicelluloses from black liquor or for mechanically fibrillating pulp. Bleaching properties are evaluated as well.

Summary

Process intensification of Nordic kraft pulp mill can be achieved by applying proposed concepts. Sulphur lean concept enables production of value-added sulphurless by-products from pulp mill and efficiency of pulp mill if gasification of black liquor is applied. An energy mill concept simplifies process by removing oxygen delignification step and by increasing the energy utilisation of pulp mill, particularly if combined heat and power production (CHP) is applied. High yield pulp concepts utilise the most valuable component wood more efficiently towards pulp products.

(219i) Sustainable and Intensified Design of a Biodiesel Production Process

Mansouri, S. S., Technical University of Denmark
Ismail, M. I. B., Technical University of Denmark
Babi, D. K., Technical University of Denmark
Huusom, J. K., Technical University of Denmark
Gani, R., Technical University of Denmark

Sustainable and intensified design of a biodiesel production process

Seyed S. Mansouri[1], Muhammad I. Ismail[1], Deenesh K. Babi[1], Jakob K.Huusom[1], Rafiqul Gani[1]

seso@kt.dtu.dk, s101298@student.dtu.dk, dkbabi@kt.dtu.dk, jkh@kt.dtu.dk, rag@kt.dtu.dk

[1]CAPEC, Department of Chemical and Biochemical Engineering, Technical University of Denmark, Building 229, SÃ¸ltofts Plads, Kongens Lyngby, Denmark

Abstract

The growing concerns about the global warming and greenhouse gas emissions (GHG) have led to an increase in the interest to produce fuel from biomass and from the fact that such fuels can relieve the reliance on imported oil and price. To this end, numerous production facilities are being set up, at different scales and using different methods of manufacture based on different raw materials and component properties [1]. It is therefore timely to study the sustainability and feasibility of these various manufacturing routes. Therefore, finding the best alternative and design with minimum environmental impact and maximum profitability is needed.

In this work a computer-aided framework for process synthesis and process intensification is applied for sustainable production of biodiesel from pure/waste palm oil as the feedstock. This approach examines several biodiesel processing routes that were collected through available data and current technologies reported in the literature. Using this information, a generic superstructure of processing routes was created that described a network of configurations representing multiple designs for the production of biodiesel. Therefore, based on the currently known technologies, this superstructure includes all possible alternatives. The next step was to analyze the superstructure in Â terms of economic and sustainability metrics. This was Â done by first performing simulation to obtain the steady state mass and energy balance data for the entire superstructure. These data were then used for a sustainability analysis [2] where a set of calculated closed- and open-path indicators were employed Â to identify the structural bottlenecks within the superstructure.

Based on this analysis the number of process alternatives within the superstructure was reduced and a set of feasible flowsheet alternatives were identified. These were further reduced through Â economic and lifecycle assessment analysis (LCA) to determine the alternative that best matched a specified set of performance criteria (or design targets). A rigorous simulation was performed on this flowsheet, which at this stage was considered as the base case design for the next step of the framework.

To further improve the base case design, process intensification was considered [3] with the targets set by LCA, economic and sustainability analyses in the previous step. Out of the three available levels for achieving PI, the phenomena-level, which is the lowest level of aggregation, was considered so that potentially new and improved alternatives to the base case design could be obtained. The objective (or target) for the intensified process design was to overcome the bottlenecks of the base case design. The optimization problem was further refined in terms of logical, operational, structural constraints, using a *PI knowledge base* tool. The next step was to identify the phenomena representing the tasks performed within the base case design and the operating window of each phenomenon, by applying thermodynamic insights [4] and the PI knowledge base. Â Next, the phenomena needed to overcome all identified process bottlenecks were identified, sorted in terms of operation (task) types and the phenomena present in them, and, screened using structural, operational and thermodynamic information. Note that different combinations of phenomena can perform the same specified task. The phenomena were then combined according to a set of rules to form unit operations, which in turn were combined to form new and innovative process alternatives. Finally, from the remaining set of feasible intensified process alternatives, the best in terms of economic and environmental sustainability was identified. For the case of biodiesel production, the intensified process alternative turned out to be the most economical and more sustainable than other alternatives.

The computer-aided methods and tools used in this work are: *SustainPro* (method and tool for process sustainability analysis), *ECON* (method and tool for process economic analysis), *LCSoft* (method and tool for process lifecycle assessment analysis) and process simulation software (e.g. PRO/II, ASPEN Plus, ICAS). They are all used in an integrated framework for process synthesis.

References:

[1] L. Simasatitkul, A. Arpornwichanop, R. Gani, ?Design methodology for bio-based processing: Biodiesel and fatty alcohol production,? *Computers and Chemical Engineering*, 2013. DOI: 10.1016/j.compchemeng.2013.01.018

[2] A. Carvalho, H.A. Matos, R. Gani, ?SustainPro-A tool for systematic process analysis, generation and evaluation of sustainable design alternatives? *Computers and Chemical Engineering*, 50, 8-27, 2013.

[3] P. Lutze, D.K. Babi, J. Woodley, R. Gani, ?A phenomena based Methodology for Process Synthesis incorporating Process Intensification,? Industrial and Engineering Chemistry Research, 2013. DOI: 10.1021/ie302513y

[4] C. Jaksland, R. Gani and K. Lien, "Separation process design and synthesis based on thermodynamic insights," *Chemical Engineering Science,* 50, 511-530, 1995.

(219h) Capture CO2 Using Diluted Aqueous Ammonia From Coal Chemicals Plant

Liu, J.,
Gao, H. C.,
Peng, C. C.,
Wong, D. S. H., National Tsing Hua University
Shen, J. F.,

After coal carbonization, a large volume of coke oven gas (COG) is produced. The COG is treated by removing corrosive and valuable components in a coal chemicals plant to provide clean flue for the iron and steel production plant. In the coal chemicals plant, diluted ammonia solutions have been used to scrub H_2S and NH_3 from COG. Therefore, in the sustainable perspective, it is more reasonable to use the existent solutions to capture the greenhouse gas, rather than using the ammine-based solutions, which need to be imported from the outside of the steel-making plant. In addition, there exist considerable low and medium pressure steams that are not reused due to the economic consideration in the steel-making plant. The temperature of regenerating the diluted ammonia absorbent is much lower than the ammine counterparts, which means that the unrecovered waste heats can be used as the regeneration energy for the CO_2 absorption process. However, the drawback of using the diluted ammonia absorbent is that the ammonia slips from the tops of the absorber and the stripper should be concerned. Additional ammonia scrubbers and an ammonia concentrator, which is used to regenerate the recycling water for the scrubbers, may be needed. The extra water and energy usages are the burden for developing the diluted ammonia process to capture CO_2. In the coal chemicals plant, there exist ammonia scrubbers removing ammonia from COG and an ammonia still process regenerating the recycling water. Therefore, if the existent processes are capable of dealing with the ammonia leakages of the CO_2 removal process, the burden of using the diluted ammonia solution can be effectively relieved.

The presented work evaluates the feasibility of integrating the CO_2 removal process with the existent processes in the coal chemicals plant. The CO_2 absorption experiment is conducted using the diluted ammonia solution from the coal chemicals plant in a laboratory pilot plant. The CO_2 removal efficiency affected by the impurities in the ammonia solution, such as: H_2S and HCN, is investigated. In addition, the absorption model is developed using the rate-based distillation module of Aspen Plus and validated by the experimental data. After that, the model is scaled up to the capacity for capturing CO_2 from blast furnace gas (BFG) in the steel-making plant. The models of the ammonia still and the ammonia scrubbers in the coal chemicals plant are also built using Aspen Plus. The feasible designs for integrating the existent processes with the CO_2 capture process are investigated.

(219j) Transesterification of Isopropyl Acetate and Methanol By Reactive Distillation

Qiu, T., Fuzhou University
Wang, H. X., Fuzhou Univerisity
Ye, C., Fuzhou University
Huang, Z., Fuzhou University

A reactive distillation process for the production of isopropanol by transesterification of isopropyl acetate with methanol is presented. The reaction has been catalyzed homogeneously by the sodium methoxide. The reaction kinetics was investigated in this work. The chemical equilibrium constant and the kinetic parameters for a homogeneous kinetic model are obtained from kinetic experimental data. Based on the fixed point analysis of the residue curve maps (RCMs), the feasibility of reactive distillation process for the transesterification of isopropyl acetate and methanol was studied. The analysis results showed that the reactive distillation process for producing isopropanol proposed in this work was feasible under certain conditions. Therefore, the reactive distillation experiments were conducted at several operation conditions (including reboiler duty, reflux ratio, space velocity, catalytic amount and reactant ratio). An equilibrium stage model of the reactive distillation was used to describe the experiments quantitatively. The simulation results are agreement with the experimental data. Subsequently, the reactive distillation model was used to determine the influence of important operating and design factors such as feed location, numbers of reactive stages and reflux ratio on reactive distillation process. On the basis of experiment and simulation studies, the appropriate process conditions were obtained. Finally, a new process for synthesis of isopropanol, combining reactive distillation with product separation columns, has been determinated.

(219l) Process Intensification: Spinning Disc Reactor Technology for TiO2 Nanoparticles Production

Mohammadi, S., Newcastle University

Boodhoo, K., Newcastle University

Harvey, A. P., Newcastle University

Process intensification: Spinning disc reactor technology for TiO_2 nanoparticles production

Somaieh Mohammadi*, Kamelia V.K. Boodhoo, Adam P. Harvey

School of Chemical Engineering & Advanced Materials,

Newcastle University, Merz Court,

Newcastle Upon Tyne NE1 7RU, UK

Corresponding author: s.mohammadi@ncl.ac.uk

Tel: +44 191 222 7169

Fax: +44 191 222 5292

Abstract

An in-depth study of TiO_2 precipitation has been performed on a spinning disc reactor (SDR). Effects of physical parameters such as rotational speed, surface texture and size, and operating parameters such as flowrate, ratio of precursor to water and location of feed introduction points were studied.

Samples collected from the disc were subjected to dynamic light scattering (DLS) to quantify the particle sizes and particle size distribution. At higher disc rotational speeds, particles with sizes close to 1 nm and with a narrow particle size distribution (PSD) were formed due to the high disc speeds, the thin films formed on the surface of SDRs experience high mixing intensity and a high degree of plug flow. Smaller particles with a narrower PSD were obtained using the grooved disc texture because the grooved disc is more effective at ensuring plug flow behaviour compared to a smooth disc under the same operating conditions.

Smaller particles and narrower particle size distributions were achieved by introducing the titanium tetra isopropoxide (TTIP) precursor into the water film away from the centre which allows better mixing between the two streams.

1. Introduction

Most pharmaceuticals and fine chemicals are produced in stirred tank reactors. The degree of heat/mass transfer and mixing of stirred vessels is reduced at larger scales because the surface to volume ratio decreases and the distribution time of contents increases, whilst the corresponding mixing intensity decreases at invariable stirrer speed. Accordingly, large vessels tend to inhibit reactions, which may be inherently rapid and highly exothermic.

The SDR is regarded as a key process intensification (PI) development in chemical production. Such PI significantly decreases the scale of the production process, which can lead to a smaller and simpler plant, lower material costs, lower waste, improved temperature control, excellent heat transfer and a safer process

[1].

The fluid residence times in a SDR are in the range of a few seconds compared with a few hours in a stirred tank. The SDR generates higher mixing intensity than a stirred vessel and should consequently be capable of retaining uniform concentration profiles within a fast reacting fluid. Therefore, better control would be applied over the reaction path than that possible in a conventional stirred vessel. Considering the pharmaceutical industry's drive to cut down production times, the SDR, with its high potential productive capacity, may allow laboratory-scale or pilot-scale vessels to achieve the same production levels as full scale vessels, thereby preventing delays in validating various levels of scale-up. Small reactor holdup (<100 mL) and tighter fluid temperature control make the SDR suitable for highly hazardous reactions [2].

The high mixing intensity within the liquid film, enhanced by waves and ripples on the interface, means that it is attractive for homogeneous reactions such as crystallisation and precipitation. An important aspect of employing SDR as the apparatus of choice, for precipitation production of nanoparticles, is the micromixing time, which needs to be very short to ensure that homogeneous nucleation dominates [3]. The micromixing [4] and macromixing [5] in SDR films have been studied recently. The SDR behaves as a plug flow reactor at a wide range of rotational speeds and flowrates [5]. A plug flow crystalliser provides excellent productive crystallisation methodology and a consistent product quality (narrow particle size distribution). PI techniques might facilitate better methods of precipitation of TiO_2 and other products, as demonstrated by the precipitation of barium sulphate on a spinning disc, which yielded significantly smaller crystals than the batch technique [6]. The main factor controlling this was the very high rates of mixing on the spinning disc, which led to the rapid depletion of supersaturation, and much higher nucleation rates. Cafiero et al. [6] also demonstrated that the energy input in the spinning disc process was much lower than the use of a T-mixer arrangement, suggesting that operating costs would also be reduced along with better control of crystal size.

Nanosized TiO_2 represents a promising research subject for various modern fields of science and technology, including nanobiotechnology and fundamental medicine [7]. Nevertheless, the synthesis and stabilisation of nanodispersed forms of TiO_2 are challenges for industry and better methods of production are required. Therefore, the intention of this study is to determine the experimental conditions for which consistent product quality for TiO2 particle formation prevails in the spinning disc reactor.

2. Materials and Methods

To characterize and predict experimentally the performance of the SDR, TiO_2 precipitation experiments were performed on a 30 cm stainless steel rotating disc. Smooth and grooved disc (8 concentric grooves) textures were studied.

The schematic set up of the rig is illustrated in Figure 1(a).

Titanium Tetra Iso Propoxide (TTIP) (the precursor) and acidified water (pH = 1.5) were introduced to the surface of the disc at 50 ^{0}C. TTIP was injected into the water film at 3 different radial positions: centre of the disc; 5 cm distance to centre and 10 cm away from the centre as shown in Figure 1(b).

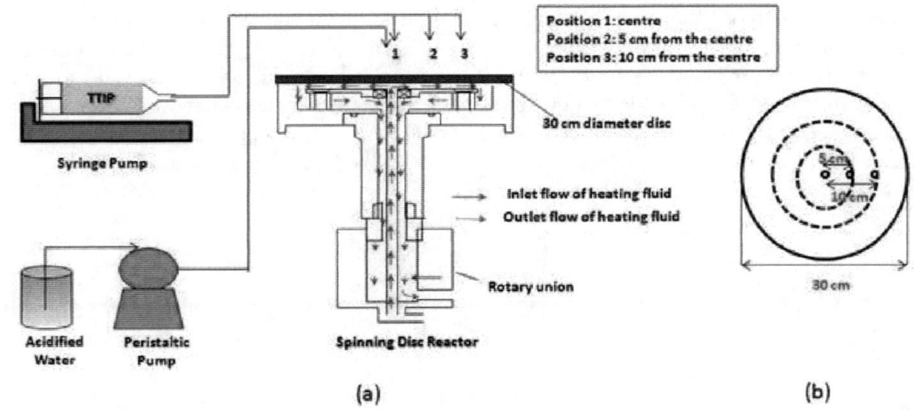

Figure 1. (a) Schematic setup of experiment, (b) top view of disc, showing TTIP stream injection positions

3. Results and Discussion

Figure 2 shows the effect of rotational speed on particle size and particle size distribution on both the smooth and grooved discs. It can be seen that at higher disc speeds that narrower size distribution and smaller particles are achieved. Higher disc speeds cause strong shearing forces and produce a thin layer of the reagent solution, which results in uniform heat transfer and homogenous concentration fields throughout the entire reaction mixture. An increase in rotational speed also increases the intensity of surface waves and promotes transverse mixing across the film thickness, thus achieving a more uniform velocity profile at any given radial position. Consequently, after nucleation, all the particles have very similar growth conditions, resulting in narrower size distribution of the nanoparticles. Figure 2 also indicates that at identical operating conditions, a grooved disc leads to production of smaller and narrower particle sizes. A grooved disc is more effective at establishing plug flow than a smooth disc. This has been attributed to continuous film detachment and reattachment on the grooved disc which enhances turbulence in the film [5].

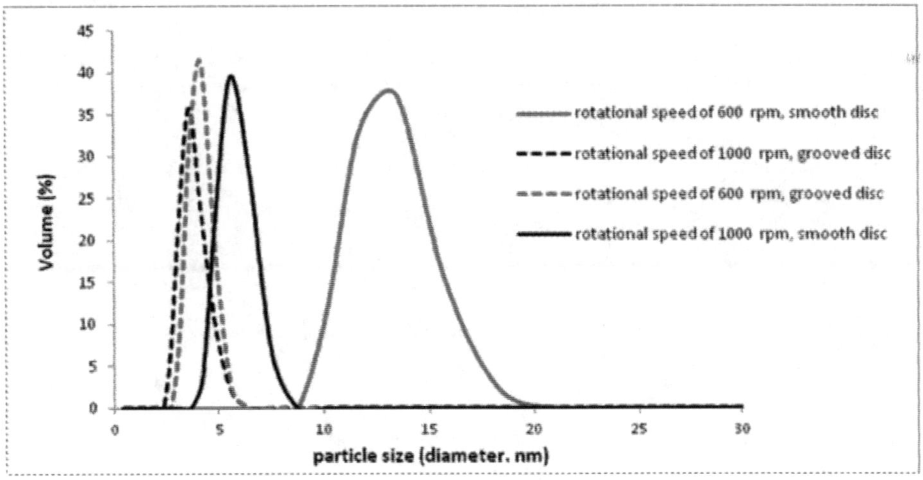

Figure 2. Effect of rotational speed and disc texture on particle size distribution

The effect of feed entry position on the particle size and distribution is shown in Figure 3. By injecting the TTIP at increasing distances from the centre, smaller particles and narrower particle sizes can be produced. It is probable that there is better mixing within the outer section of the disc because of increased shear rate as the film thins out towards the edges. It is expected that nucleation rates were higher and more uniform throughout the film in this highly mixed environment, giving smaller particles with narrower PSDs. Moreover, with injection further way from the centre, the residence time of the mixed fluid stream is shorter. This is also likely to contribute to reducing the extent of growth of the particles, resulting in smaller particles.

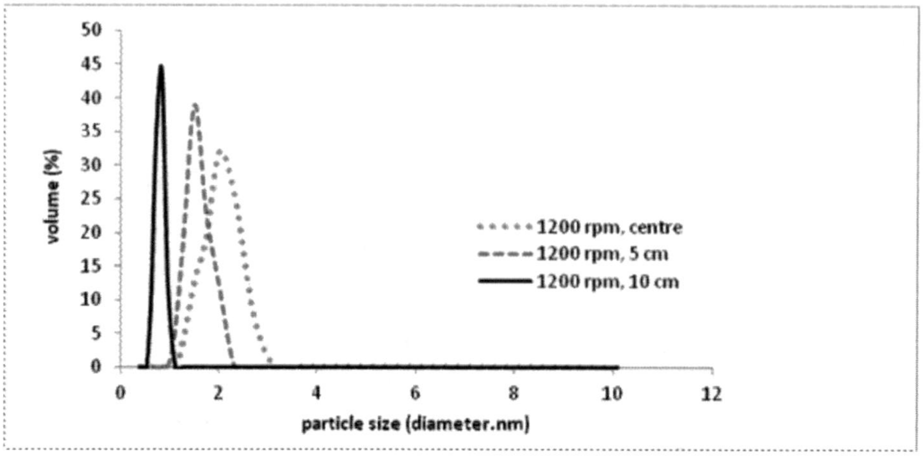

Figure 3. Effect of feed location on particle size distribution on smooth disc

4. Conclusion

In this work, it has been shown that the intensified processing of TiO_2 precipitation under high centrifugal fields on the spinning disc reactor offers advantages as, in the ciorrect conditions, production of very small (1-2 nm)particles with a very narrow PSD is possible.

At a rotational speed of 1200 rpm on a 30 cm diameter smooth disc with the feed pipe (feeding TTIP) 10cm from the centre, particles of below 1 nm diameter were obtained. Under the same conditions the grooved disc produced even smaller particle sizes than the smooth disc and with narrower distribution.

This intensified method of production of TiO_2 particles represents a viable way forward to continuous, industrial production of such an important chemical.

5. References

1.
Reay, D., Ramshaw, C., Harvey, A., *Process Intensification, engineering for efficiency, sustainability and flexibility*. 2008: Butterworth-Heinemann Ltd

2.
Pask, S.D., O. Nuyken, and Z.Z. Cai, *The spinning disk reactor: an example of a process intensification technology for polymers and particles.* Polymer Chemistry, 2012. **3**(10): p. 2698-2707.

3.

Nielson, A.E., *Homogeneous Nucleation in Barium Sulfate Precipitation.* Acta Chemica Scandinavica, 1961. **15**: p. 441-442.

4.
Boodhoo, K.V.K. and S.R. Al-Hengari, *Micromixing Characteristics in a Small-Scale Spinning Disk Reactor.* Chemical Engineering & Technology, 2012. **35**(7): p. 1229-1237.

5.
Mohammadi, S. and K.V.K. Boodhoo, *Online conductivity measurement of residence time distribution of thin film flow in the spinning disc reactor.* Chemical Engineering Journal, 2012. **207**: p. 885-894.

6.
Cafiero, L.M., et al., *Process intensification: Precipitation of barium sulfate using a spinning disk reactor.* Industrial & Engineering Chemistry Research, 2002. **41**(21): p. 5240-5246.

7.
Chen, X. and S.S. Mao, *Titanium dioxide nanomaterials: Synthesis, properties, modifications, and applications.* Chemical Reviews, 2007. **107**(7): p. 2891-2959.

(219e) Applications of Flowscan^{TM} Non-Invasive Rheological Measurement of Foods and Personal Care Products

Tozzi, E. J., University of California, Davis
McCarthy, M. J., University of Calfiornia Davis

Maintaining strict rheological parameters in the manufacture of liquid foods and personal care products is essential for ensuring product quality, manufacturing efficiencies, acceptance testing and consistent process control. Adequately controlling rheological properties in batch or continuous processes with manual testing methods can be challenging due to slow feedback loops related to the extraction, preparation and measurement of samples. Moreover, rheological properties can change during sample handling and storage, thereby affecting the final results. In sensitive processes, the lack of tight rheological control can result in high variability in product, excessive use of ingredients, wastage, and product rejections. Automated non-invasive rheology measurements of in-line product flows, on the other hand, provide real-time quantification of process conditions, which simplifies process monitoring, process control and detection of anomalies. Early detection of a process deviation can result in significant savings of time, materials and overall costs. To demonstrate these capabilities, we used the FlowScan™ real-time rheology system (Aspect Imaging) in a pipe flow system with 30 liter capacity to dynamically monitor the rheology of various personal care products as well as dairy products undergoing rapid rheological changes. The capabilities of FlowScan to measure other properties of these liquids in addition to rheology are discussed.

(219f) Energy Efficient Control of Distillation Process

Osuolale, F. N., Newcastle University
Zhang, J.,

Improving the energy efficiency of new and existing processes is an increasingly important issue for consideration by industrialist, academia and policy makers. This is because of the economic, environmental and ecological implications of unutilised energy. Distillation is one of the most widely used energy intensive separation technique in the chemical and petrochemical industries. Distillation unit however poses a great challenge to control engineers because of its complexity, specific operating objectives, significant interactions among the control loops and specialised constraints. These have often led to trade-off between energy usage and product quality in distillation column control and operations. Several methods of control configuration selections are available in the literatures; however, few studies have been reported concerning the energy efficiency of various control configurations. This study details the choice of an energy efficient control configuration by incorporating thermodynamics second law in the selection criteria. Two binary systems of methanol-water and benzene-toluene were considered. Three control configurations selected for each of the system are LV, LB and DV. Relative gain array (RGA) and relative exergy array (REA) were used in screening for the best control configuration for each system. REA incorporates minimum energy usage in the control structure selection. A full detailed analysis of the performance of the controlled structure was conducted in the dynamic mode. Controller parameters were tuned using the biggest log modulus (BLT) method. Set point change in the distillate and bottoms composition were made for each of the configuration and their closed loop responses were obtained. Exergy loss, exergy efficiency and energy usage for the dynamic simulations were then calculated. The results show that the control configuration that was selected as the best from a combination of the RGA and REA for each of the system also has the least energy usage and exergy loss in the dynamic simulation. Thus the process operability analysis results are supported by simulation results. Incorporating energy efficiency in control structure selection will ensure the control of a distillation process not only in terms of product quality and safety but for energy efficient operations as well.

(219m) A Design Tool for Microchannel Heat Exchangers

Kim, Y., Korea Institute of Machinery and Materials
Yoon, S. H.,
Oh, D. W., Korea Institute of Machinery and Materials
Choi, J., Korea Institute of Machinery and Materials
Lee, K. H., Korea Institute of Machinery and Materials

Microchannel heat exchangers have potentials to advance various thermal and chemical processes by extending operating range or improving system performance. In some applications such as a very high pressure system, diffusion-bonded microchannel heat exchangers are the only option available to endure such harsh condition.

A design tool has been developed to facilitate optimal design of microchannel heat exchangers. Microchannel heat exchangers are designed and manufactured by chemical etching and diffusion bonding. Their cryogenic performance is investigated in a demo scale test facility. Empirical correlations derived from a set of experiments are embedded to the software.

(219k) Transesterificatiom of Jatropha Oil in the Presence of Various Co-Solvents

Goyal, R., Indian Institute of Technology Roorkee
Kumar, D. V., IIT Roorkee
Mishra, I. M., Indian Institute of Technology Roorkee

Fatty acid methyl esters (FAME), also known as biodiesel, derived from renewable feedstocks which fall in the carbon range C12-C22, and have similar properties as diesel. Typically the biodiesel is produced with transeterification reaction, where vegetable oil is reacted with methanol in the presence of catalysts. The biodiesel yield is affected by various factors: type of alcohol, molar ratio of alcohol to vegetable oil, type of catalyst (acid or basic), amount of catalyst, reaction temperature, reaction time and feedstock quality (e.g. free fatty acid content, water content etc.). In the present work the fatty acid methyl esters are synthesized from non-edible feedstock (jatropha oil) of indian origin. The synthesis of FAME from jatropha oil is carried out in two-step process i.e. acid catalysed esterification followed by base catalysed transesterification for the production of biodiesel. The transesterification of jatropha oil is carried out using various co-solvents: diethylether (DEE), tert-butyl methyl ether (tBME), and tetrahydrofuran (THF). The catalyst used was sodium methoxide (CH3ONA). The study is carried out to find the optimal operating parameters for jatropha oils in batch reactor. The transesterification reaction is carried out for the following conditions for different solvents: 1.0 wt% of CH3ONa, a molar ratio methanol/oil of 9:1, a molar ratio co-solvent/methanol 1:1, an agitation rate of 700 rpm, and a temperature of 30 °C. The result of the experimental study shows that the yield of triglyceride into ester increases with increase in time. It has been observed that the approximately 85% reaction completes in 30 sec. The qualitative analysis of vegetable oils is carried out using thin layer chromatography (TLC). DEE and THF are found as the best solvent as compared to the other solvents used in the present work. A similar yield is obtained with both DEE and THF, however separation process is difficult and takes longer time as compared to DEE. Presently the studies are going on to characterize the fuel properties (e.g., density, viscosity, flash point, fire point, cloud and pour point, diesel index, water and free fatty acid contents) of biodiesel.

(219c) Analysis of Full Scale Fixed Bed Adsorption of Active Pharmaceutical Ingredient By Applying a Validated Adsorption Process Model Based On Independently Determined Key Adsorption Parameters

Senica, D., Delft University of Technology
Likozar, B., National Institute of Chemistry
Pavko, A., University of Ljubljana

Adsorption is a common separation method for isolation and purification of active pharmaceutical ingredients (APIs) produced by bio-syntheses, but despite its widespread use, published literature describing such adsorption processes, their mathematical modeling, and the use of models for design, scale-up, optimization and troubleshooting is rare. Therefore a mathematical model (a simple non-structural and more complex structural approach was used) for a continuous fixed bed column adsorption process of glycopeptide antibiotic vancomycin was developed as a tool for analysis of the operation of a full scale adsorber. The model allows a prediction of breakthrough curves by an application of adsorption kinetics and equilibrium data, determined independently in batch adsorption experiments. Parameters characterizing the flow pattern of the liquid in the adsorption column, determined independently as well, using a tracer method. Such an approach differs from a widespread practice of fitting experimental breakthrough data and lumping several parameters into a dispersion coefficient.

The model was validated with laboratory scale experiments. Two macro-reticular polymeric adsorbents (Rohm&Haas XAD16N and XAD1600N), differing in particle and pore size distribution, were evaluated and their vancomycin dynamic adsorption capacity was determined for different superficial velocities. The key model parameters, such as those related to adsorption equilibrium and kinetics, were determined separately in batch experiments. A description of the liquid flow pattern in the column was obtained on the basis of tracer experiments, while different correlations have been utilized for the estimation of liquid phase mass transfer coefficients. Comparison of modeled and experimental break through curves shown that virtually all the result fell within the ±25% margin, typically even within ±15%. A certain amount of a disagreement between the model-predicted and the measured data may primarily be the result of the non-ideal wetting of adsorbent particles and various channeling effects, which render the adsorption kinetics, determined by batch experiments, less realistic; nonetheless, within a certain margin of error, our approach with data based on batch experiments was shown to be adequate. Furthermore, the parametric sensitivity of the structural adsorption/desorption kinetics model was evaluated to determine the impact of changes in key model parameter on the model results.

Laboratory scale experiments confirmed that the use of Amberlite XAD1600N is advantageous, which, based on the batch adsorption data, was expected. The dynamic adsorbent capacity (determined at C/C_0 = 0.1) was approximately 2.5–3.0 times higher (50–90 g kg^{-1} wet resin) compared to Amberlite XAD16N

(20−35 g kg^{-1} wet resin). The predicted values of the dynamic capacity were somewhat underestimated, but typically within the +10/−35% range. As expected, lower liquid velocity (*i.e.*longer residence time) was favorable for the dynamic adsorption capacity.

Finally, the validated model was applied for an analysis of the existing full scale adsorption process and predicted break-through curves revealed 20-30% of unused adsorbent capacity and therefore the column can be further loaded with no loss of API. To increase the capacity of the adsorption process even more, adsorbent has be replaced either with fresh XAD16N or with XAD1600N.

The laboratory scale model verification and the application of the model to analyze the operating boundaries of the existing adsorption process demonstrated that our methodology in modeling the column adsorption process based on independently determined key model parameters, assures a reliable prediction of break through curves and dynamic adsorption capacities. The approach therefore can be applied as a valuable tool for fast evaluation of the effectiveness of various adsorbents, experimental design, designing full-scale continuous column operation, scaling-up, and analysis and troubleshooting of the existing full-scale adsorption processes.

(219b) The Mesoscale Oscillatory Baffled Reactor as a Robust Intensification Reactor for Rapid Process Development

Eze, V. C., Newcastle University
Harvey, A. P., Newcastle University
Phan, A. N., Newcastle University
Fisher, J. C., Newcastle University

Multi-dimensional, dynamic and continuous steady state screening of an acid-catalysed esterification reaction were carried out using a mesoscale oscillatory baffled reactor ("meso-OBR") to investigate the ability of the reactor for solid-liquid reactions. Sulphonic acid functionalised Amberlyst (Amberlyst^{TM} 70) was used as a catalyst for the reaction at a temperature of 60 °C under an oscillation condition (frequency of 4.5Hz and amplitude of 8mm) at methanol to hexanoic acid molar ratios of 1:1 – 30:1 and residence times ranging from 1min to 20min.. Steady states were achieved at all the residence times used for the continuous multi-steady states and all the ramped feed molar ratios. The hexanoic acid conversions increased with residence time from 54.98 ± 0.86% at 2.5min through 80.05 ± 1.39% at 10min to 95.43 ± 1.04% at 20min residence times and 30:1 methanol to hexanoic acid molar ratio. The Amberlyst^{TM} 70 recovered its catalytic activity fully after the system was spiked with water, indicating that the catalyst does not suffer permanent deactivation in the presence of water. The hexanoic acid conversions were similar and reproducible for the multi-dimensional, dynamic and continuous steady states. These results showed that rapid process development can be achieved using the meso-OBR in multi-dimensional and dynamic modes.

(230a) Chromatographic Scale-Up On a Volume Basis

Hansen, E. B., Novo Nordisk

Originally presented on: 11/5/2013 8:30:00 - 8:55:00

The scale-up of a chromatographic separation is in principle a simple procedure since the process parameters are scalable in a linear fashion. The process is scaled by increasing the column diameter while keeping the bed height, the velocity, and the volumes of the different phases (measured in column volumes) constant. The problem with this simple and well established concept, usually referred to as linear scale-up, is the lack of flexibility.

For practioners, especially those dealing with large scale operations, it is no secret that chromatography columns come in discrete sizes regarding the diameter. The diameter and the number of columns are the only design variables we have available when going into production scale. As a consequence we most often cannot choose the column volume that matches the desired capacity. We are faced with the problem that for a chosen diameter we need for instance 2.2 columns in order to meet the capacity demands. The solution is either to invest in a third column or go back to the lab and try to optimize the process in order to squeeze it into two columns. Both solutions can be rather costly either in investment or in time consumption in development. Further, we have no guarantee that the optimization will actually succeed (not to speak of the documentation burden if it does). Even when we manage to meet the right capacity we still risk encountering the same problem at a later date if we want to increase the production capacity. In pilot scale the problem can be even more pronounced since we must use existing equipment. The lack of flexibility of the column volume can therefore increase the number of cycles on each column and thereby the duration of the pilot campaign (where the chromatography steps are often the bottleneck). At the other end of the scale, in the development phase, we are met with similar problems. Knowing that the process will end up in production some day and the bed height must be kept constant we aim for a suitable bed height (like 20 cm) when developing the chromatography in lab scale. In order to avoid packing problems and flow irregularities a certain column diameter (~ 1 cm) is required and we therefore end up with a lower volume limit around perhaps 15-20 mL in order to have a representative process step. Especially in the early phase of development where the product quantities are limited this can be a substantial volume which reduces the number of trials and we therefore risk a delay in development. A volume in this range is also certainly not desirable for special tasks like virus validation.

Fifteen years ago we revisited the scaling principles of chromatography in order to find a more flexible way of scaling up (or down). From a conceptual point of view the solution was quite simple and does not involve any new theory. By scaling the process based on a constant contact time (residence time) we can, within limits, introduce the column bed height as a process design variable rather than a fixed parameter. Since the bed height is a continuous variable, as opposed to the diameter, we can use it to obtain exactly the volume we need in any scale. After proof of concept in lab scale and later in pilot scale the new concept became an established approach which has proven extremely valuable over the entire range from development in lab scale to commercial production. We have previously presented and published some of our early results. Our expectation at the time was that this would become an established approach at least for pharma companies producing significant volumes. But our experience from interactions with companies, including producers of chromatography equipment, and with the regulatory authorities has shown that this is apparently still new territory. Since we find the approach very useful to the field of industrial chromatography we would like to take the opportunity to present ten years of experience in chromatographic scale up on a volume basis. The presentation will include the theoretical background and proof of concept as well as industrial cases with scaling factors of more than 1000. Examples include IEX, HIC, RPC, and SEC processes for the separation of both peptides and large proteins.

(230b) Measurement, Prediction, and Control of the Physical Stability of a Hygroscopic API

Boehmcke, P., Bristol-Myers Squibb
Domagalski, N., Bristol-Myers Squibb Company
Tabora, J. E., Bristol-Myers Squibb
Originally presented on: 11/5/2013 8:55:00 - 9:20:00

The selection of hygroscopic crystalline forms is sometimes necessary to enable the clinical development of API. In this work the drug substance isolated from the manufacturing process is isolated as an anhydrous crystalline form which readily converts to a hydrate under environments of moderate relative humidity. Form control during storage and drug product manufacture was necessary to ensure process robustness. This presentation will focus upon experimental and modeling studies to enable definition of a control strategy which preserves the anhydrous form of the API. This strategy was applied to packaging and storage of the drug substance as well as processing and sampling of the drug product.

The talk will discuss a experimental set up designed to measure the dynamic hydration under different relative humidity environments by automatically recording the relative humidity, temperature, and sample mass. Different geometries were used to examine the impact of cake length on hydration rate.

Finally, a model incorporating solid state reaction kinetics and powder vapor diffusion was developed from the experimental data. The model allowed for a quantitative assessment of the risk and impact of hydration under different manufacturing and storage environments.

(230c) The Application of Advanced Control and Model Based Strategies to Biopharmaceutical and Pharmaceutical Processes

Barrett, M., University College Dublin
Whelan, J., University College Dublin
Carr, M., APC Ltd
Glennon, B., UCD
Originally presented on: 11/5/2013 9:20:00 - 9:45:00

Pressures are increasing to develop, optimise and scale-up processes in minimal time frames whilst maximizing manufacturing related process understanding (QbD). In turn strategies such as process control and modelling are emerging as powerful platforms to compliment and maximise the output from current experimental programs.

Presented are novel model based control strategies that enable the streamlined development, scale-up and scale-down of bio/pharmaceutical unit operations. Cell culture and crystallization examples will be shown where the interrogation of these process development platforms has facilitated a fundamental understanding of the systems kinetic trajectory and environmental sensitivities. This process insight enabling better process titre, reliability, scale-up (cell culture) and particle shape, size and control (crystallization) to be delivered across development and manufacturing scales.

These platforms importantly combine both scale (i.e. mixing) and kinetic (i.e. secondary nucleation, decreasing product titre) related effects to facilitate the development of manufacturing based strategies within a laboratory environment. The introduction of 'virtual lab and plant' aspects of these platforms will also be presented whereby the predictive nature of these tools aids process design for scale-up / scale-down and tech transfer.

(230d) Modeling of the Secondary Drying and the Fate of Organic Solvents for Spray Dried Dispersion (SDD) Product

Hsieh, D., Bristol-Myers Squibb Company
Yue, H.,
Nicholson, S.,
Roberts, D., Bristol-Myers Squibb Co.
Schild, R., Bristol-Myers Squibb
Gamble, J., Bristol-Myers Squibb
Lindrud, M., Bristol-Myers Squibb Co.
Originally presented on: 11/5/2013 9:45:00 - 10:10:00

The spray dryer has been used to generate amorphous dispersions of API and polymer to enhance the solubility and bioavailability of drugs that are less water soluble. This powder, after spray drying, contains residual solvents and they need to be further removed to an acceptable limit in a secondary dryer. The purpose of this study is to understand the secondary drying mechanism and establish a model to predict the drying performance for scale-up and the fate of organic solvents.

The kinetics of solvent removal from Spray Dried Dispersion (SDD) product follows the Fickian diffusion behavior. Since the SDD particles are typically hollow and the shell thickness of the hollow bead is typically between 10-15% of the diameter of the bead, the effect of curvature can be neglected. Hence, a diffusion equation for the geometry of a slab as an approximation was used for modeling the secondary drying. Excellent agreement is obtained between the experimental data and the prediction from the modeling. The control parameters of the secondary drying are the diffusion coefficient and the diffusion length. Since the diffusion coefficient is a function of temperature only, the secondary drying performance is not scale dependent. This secondary drying has been successfully scaled up from a lab scale of a few grams to a bench scale of 500 grams in an agitated conical dryer and the large-scale drying kinetics follow the model prediction well. The fate of residual solvents including acetone, methanol and others can be successfully determined from this modeling work.

(230e) Complex Fitness Diagrams and Downstream Processing

Lightfoot, E. N., University of Wisconsin--Madison

Originally presented on: 11/5/2013 10:10:00 - 10:35:00

Focus here is development of production scale processes and specifically with modifying optimized laboratory processes that are unsuitable for large-scale operations. Our example is production of high-value fermentation produced biologicals such as therapeutic proteins, and we approximate value in terms of evolutionary fitness dagrams. We also suggest that speed and economy of development can often be ncreased quite significantly by judicious incorporation of heuristics. We show that one is rapidly led bythis kind of thinking to radical but highly beneficial new processes ffor industrial scale operation. Success un such a procedure rquires careful specification of system and goals.

(230f) Bilayer Tablet Compaction: Translation From a Single Stage Compaction Simulator to a Rotary Press

Rowe, J. M., Bristol-Myers Squibb
LaMarche, K., Bristol-Myers Squibb Company
Carroll, B., ExxonMobil
Pandey, P., Bristol-Myers Squibb
Subramanian, G. A., Bristol-Myers Squibb
Nikfar, F., Bristol-Myers Squibb
Originally presented on: 11/5/2013 10:35:00 - 11:00:00

Fixed dose combination products (FDC) are formulations containing two or more active pharmaceutical ingredients (API) combined within a unit dose. In recent years, FDCs have become increasingly popular in several therapeutic areas because they offer convenience, a reduced pill burden, and cost savings for patients.

In cases when the APIs are chemically incompatible or multiple release profiles are desired, the dose must be formulated as a layered tablet. However, layered tablets pose unique development and manufacturing risks because of the fine balance required between maintaining sufficient physical separation between the layers to mitigate chemical incompatibilities and ensuring that the tablet remains intact (i.e. avoid layer delamination) throughout the manufacturing process and product shelf life. Challenges during the manufacture of layered tablets can be difficult to resolve due to the limited mechanistic understanding of the compaction process. Thus, the development process for layered tablet formulations has been largely trial-and-error, requiring significant resource investments, especially when considering that layered tablet production on a rotary press requires kilogram-quantities of materials.

In this work, the authors have sought to gain mechanistic understanding of the compaction process for layered tablets to minimize resources required during product development, more specifically to better understand how experimental factors (i.e. material selection, compaction conditions) impact layer interface strength. Since the failure of layered tablets typically occurs due to delamination at the interface, a fundamental appreciation of the governing mechanisms that impact interface strength is crucial in order to improve the overall mechanical integrity of that layered tablet. A single station compaction simulator, fitted with a three-sensor instrumented die, was used as a material-sparing tool (requires only grams of material) for the development of bilayer tablets and to increase fundamental understanding of the compaction process. The instrumented die was used to measure the die-wall pressure exerted by each layer of the bilayer tablet during the compaction process.

Bilayer combinations of placebo and active granulations over a range of compaction forces were examined in this study. Young's moduli and Poisson's ratios, as well as additional elastic material properties described by the Drucker-Prager-Cap model, were determined using measured die-wall pressure profiles for each material in the study. Analysis of the die-wall pressure profiles during bilayer compaction yielded a correlation between the residual stress within each of the layers after compaction and the interface strength of the bilayer tablet. The level of residual stress experienced by different materials, when compressed as either the first or second layer in a bilayer tablet, was related back to elastic material properties. Using this knowledge, select material combinations were compressed as bilayer tablets on a rotary tablet press. The trends in tablet interface strength for tablets produced on the compaction simulator were similar for those manufactured on the rotary press. However, it was

noted that for certain material combinations, tablets made on the compaction simulator and the rotary press did

not possess the same solid fraction when compacted under the same forces. When tablet strength was examined

as a function of solid fraction instead of compaction force, the compaction simulator was found to adequately

emulate the rotary press. Discrepancies between press compaction force and corresponding solid fractions were

related back to elastic material properties. In conclusion, the compaction simulator was found to be a useful

tool to develop bilayer tablet formulations at small scales, providing results that may be translated to larger scale

production on a rotary press.

(241a) Mathematical Modeling and Design of Layer Crystallization in a Concentric Annulus With and Without Recirculation

Zhou, L., Massachusetts Institute of Technology
Su, M., Massachusetts Institute of Technology
Benyahia, B., Massachusetts Institute of Technology
Singh, A., Massachusetts Institute of Technology
Barton, P. I., Massachusetts Institute of Technology
Trout, B. L., Massachusetts Institute of Technology
Myerson, A. S., Massachusetts Institute of Technology
Braatz, R. D., Massachusetts Institute of Technology
Originally presented on: 11/5/2013 8:30:00 - 8:52:00

In pharmaceutical manufacturing, the drug product is often separated from a liquid phase by crystallization, washing, filtration, and drying—each of which reduce the overall product yield. Poorly filtering crystals can result in bottlenecks in the downstream processing and can affect crystal product purity or reduce yield due to additional washes. These observations have motivated efforts to design continuous crystallization processes that are reliable for manufacturing a product that meets purity, yield, shape, and size requirements, and at the same time ensures efficient downstream processing.

One way to avoid slurry handling is to constrain crystal growth to occur on fixed surfaces so that the amount of solid particles in the bulk is minimized. Layer crystallization has been characterized as being a process in which coherent crystal layers grow on cooled surfaces of a specially designed tube or plate heat exchanger.[1,2] Layer crystallization from melts has been widely applied for the purification of organics and inorganics such as sucrose and milk fat, and in seawater desalination.[3-6] For layer crystallization, the liquid phase is always at a higher temperature than the solid phase, so that in melts, the temperature is required to be higher than the melting point, but layer crystallization from solutions can be operated at much lower temperatures with less energy consumption.[7]

This work proposes a layer crystallization process from a solution in a concentric annulus that avoids the need for washing, filtration, and drying.[8] A continuous flow of drug product could be generated by employing multiple cyclic annular crystallizers in parallel. Many past studies on mass[9-12] and heat transfer[12,13] in annular reactors have concentrated on obtaining correlations of dimensionless number, but these methods are limited by the configurations and operating conditions[13,14] and they are not able to predict the entire concentration or temperature field that is important for crystallization. To our knowledge, this work for the first time thoroughly examines heat and mass transfer of a crystallization process with solid-layer growth in an annulus with and without recirculation. The mass and/or heat transfer model with solid-layer growth on a cylindrical surface can be employed for many other applications, such as fouling in heat exchangers,[15] biofilms in petroleum pipelines,[16] and melt layer crystallization in a tank.[17,18]

A dynamic model is developed in the form of coupled partial differential equations to (1) predict the crystal layer thickness as a function of time and axial position, (2) update the nonuniform boundary temperature at the solid-liquid interface, which is affected by the local thickness of the crystal layer, and (3) calculate the temperature and concentration fields in the annulus. The model predictions are shown to closely track experimental data that were not used in the model's construction, and also compared to an analytical solution that can be used for quickly obtaining rough estimates when there is no recirculation loop. The effects of modifying the operating conditions are investigated, including variation in initial solute concentrations, cooling liquid jacket temperature, inlet solution velocities, core liquid temperature, and extent of mixing in the recirculation loop. The model can be used to optimize product yield and crystal layer thickness uniformity, with constraints on the supersaturation to avoid bulk nucleation by adjusting cooling temperatures in the core and jacket. Additionally, the model could be extended to predict molecular purity of the crystal layer, by considering more components.

References

1. Glade H, Ulrich J. *Melt Crystallization: Fundamentals, Equipment and Applications*. Aachen, Germany: Shaker; 2003.

2. Ulrich J, Bülau HC. 7 - Melt crystallization. In: Myerson AS, ed. *Handbook of Industrial Crystallization*. 2nd ed. Woburn: Butterworth-Heinemann; 2002:161-179.

3. Parisi M, Chianese A. The crystal layer growth from a well-mixed melt. *Chemical Engineering Science.* 2001;56(14):4245-4256.

4. Rich A, Mandri Y, Mangin D, Rivoire A, Abderafi S, Bebon C, Semlali N, Klein J-P, Bounahmidi T, Bouhaouss A, Veesler S. Sea water desalination by dynamic layer melt crystallization: Parametric study of the freezing and sweating steps. *Journal of Crystal Growth.* 2012;342(1):110-116.

5. Flesland O. Freeze concentration by layer crystallization. *Drying Technol.* 1995;13(8-9):1713-1739.

6. Peters-Erjawetz S, Ulrich J, Tiedtke M, Hartel R. Milk fat fractionation by solid-layer melt crystallization. *Journal of the American Oil Chemists' Society.* 1999;76(5):579-584.

7. Aumock NM. *A Process for Separation by Semi-Continuous Counter-Current Crystallization*. Cambridge, MA: Dept. of Chemical Engineering, Massachusetts Institute of Technology; 2011.

8. Zhou L, Su M, Benyahia B, Singh A, Barton PI, Trout BL, Myerson AS, Braatz RD. Mathematical modeling and design of layer crystallization in a concentric annulus with and without recirculation. *AIChE Journal.* 2013;59(4):1308-1321.

9. Ross TK, Wragg AA. Electrochemical mass transfer studies in annuli. *Electrochim. Acta.* 1965;10(11):1093-1106.

10. Mobarak AA, Farag HA, H.Sedahmed G. Mass transfer in smooth and rough annular ducts under developing flow conditions. *J. Appl. Electrochem.* 1997;27(2):201-207.

11. Ould-Rouis M, Salem A, Legrand J, Nouar C. Etude numÃ©rique et expÃ©rimentale des transferts de matière et de quantitÃ© de mouvement dans un Ã©coulement annulaire laminaire non Ã©tabli. *Int. J. Heat Mass Transfer.* 1995;38(6):953-967.

12. Ghosh UK, Upadhyay SN. Mass transfer to newtonian and non-newtonian fluids in short annuli. *AIChE Journal.* 1985;31(10):1721-1724.

13. Viskanta R. Heat transfer with laminar flow in a concentric annulus with prescribed wall temperatures. *Appl. Sci. Res.* 1964;12(6):463-476.

14. Mitrović J, Maletić B. Effect of thermal asymmetry on heat transfer in a laminar annular flow. *Chemical Engineering & Technology.* 2005;28(10):1144-1150.

15. Kirkpatrick JP, McIntire LV, Characklis WG. Mass and heat transfer in a circular tube with biofouling. *Water Res.* 1980;14(2):117-127.

16. Al-Darbi MM, Agha KQ, Islam MR. Estimation of the substrate concentration profiles in petroleum pipelines containing biofilms. *Pet. Sci. Technol.* 2006;24(11):1249-1265.

17. Chianese A, Santilli N. Modelling of the solid layer growth from melt crystallization—the integral formulation approach. *Chemical Engineering Science.* 1998;53(1):107-111.

18. Guardani R, Neiro SMS, Bülau H, Ulrich J. Experimental comparison and simulation of static and dynamic solid layer melt crystallization. *Chemical Engineering Science.* 2001;56(7):2371-2379.

(241b) Cocrystals: Latest Advances in 'in Silico' Methods for Predicting Cocrystal Coformers and Their Properties

Gallagher, D. A., CAChe Research
Loschen, C.,
Originally presented on: 11/5/2013 8:52:00 - 9:14:00

The formulation, design, and implementation of active pharmaceutical ingredients (API) is an area of great interest. Changing the structure and composition of an API by cocrystallizing it with a coformer can have a significant influence on the properties and bioavailability of the drug, and can even offer the potential for new patents and associated revenues. For example, changing the solubility can influence the formation of crystals, cocrystals or solvates, as well as the pharmacokinetics through dissolution during drug release.

The traditional experimental approach to screening for possible new coformers and measuring their properties is time-consuming and expensive. However, *in silico* methods are potentially orders of magnitude faster and are becoming more accurate, such that it is now possible to prescreen huge libraries of potential coformers in minutes or hours on a laptop computer, to determine the most likely candidates. This enables the chemical engineer to take into account a much wider range of possible candidates and then focus the experimental work on the most promising ones, so greatly improving the chances of success in the lab.

This presentation will review the capabilities and limitations of some of the available *in silico* tools for predicting cocrystal coformers and their properties, and compare them with experiment, including group contribution methods, quantum chemistry, and statistical thermodynamics. In addition, the latest algorithmic and method advances for improving the prediction of compounds that can form cocrystals and their properties in a range of solvents, will be discussed.

(241c) Determination of Critical Seed Loading From Nucleation Kinetics

Tseng, Y. T., National Taiwan University
Ward, J. D., National Taiwan University
Originally presented on: 11/5/2013 9:14:00 - 9:36:00

It has been observed experimentally that in at least some cases, secondary nucleation in batch crystallization can be suppressed by using an adequate seed loading. Kubota and coworkers[1–3] call this seed loading the critical seed loading and propose a correlation to predict the critical seed loading as a function of seed size. So far, critical seed loading has only been determined experimentally for a limited number of systems. In this contribution, a method is presented for preparing the critical seed loading diagram knowing only the nucleation kinetics for the system in question. This is advantageous because it is much less time-consuming than conducting experiments or developing and solving a complete batch process model including solubility expressions and growth and nucleation rate kinetics.

The results for 43 systems taken from the literature show that there is a great deal of variability in the critical seed loading, and a single correlation is inadequate to describe the relationship between critical seed loading and seed size. For a seed size of 10 microns, seed loading can effectively suppress nucleation in 92% of cases, while for seeds of size 50, 100 and 200 microns, seed loading can suppress nucleation in 74%, 61% and 39% of cases respectively.

Furthermore, the effects of changing process parameters (total batch time and the net crystal yield) can also be readily studied, although they are found to have a small effect compared to seed size. Finally the possibility of using a critical seed length or critical seed surface area for process development instead of critical seed mass is also investigated, however these alternatives are not found to be more attractive than the critical seed mass.

References

1. Doki N, Kubota N, Yokota M, Chianese A. Determination of critical seed loading ratio for the production of crystals of uni-modal size distribution in batch cooling crystallization of potassium alum. *J Chem Eng Jpn.* Jul 2002;35(7):670-676.
2. Doki N, Kubota N, Sato A, Yokota M, Hamada O, Masumi F. Scaleup experiments on seeded batch cooling crystallization of potassium alum. *AIChE J.* Dec 1999;45(12):2527-2533.
3. Jagadesh D, Kubota N, Yokota M, Sato A, Tavare NS. Large and mono-sized product crystals from natural cooling mode batch crystallizer. *J Chem Eng Jpn.* Oct 1996;29(5):865-873.

(241d) Design and Optimization of Continuous Crystallization Processes: Cascades of Mixed Suspension Mixed Product Removal Crystallizers

Vetter, T., ETH Zurich
Doherty, M. F., University of California, Santa Barbara
Originally presented on: 11/5/2013 9:36:00 - 9:58:00

Crystallization processes in the pharmaceutical industry have been carried out for decades as batch processes. It is recognized that this type of operation suffers from product variability from batch-to-batch and comparably high manufacturing costs [1]. While continuous processing is a proven technique in many large scale industries for overcoming both these deficiencies, the pharmaceutical industry was reluctant to adopt such an approach in the rapid growth era of the twentieth century. However, competition and regulatory demands are ever-increasing and there is now more focus on reducing manufacturing costs while maintaining high product quality.

While the design and optimization methodology for batch crystallization processes are relatively well understood for various combinations of cooling, anti-solvent or reactive crystallization [2] and for different optimization objectives [3], there are yet relatively few studies targeted on the design and optimization of continuous crystallization processes that keep the specific challenges posed to the pharmaceutical industry in mind [4,5].

In this work we present a comprehensive investigation of cascades of mixed suspension mixed product removal (MSMPR) crystallizers based on population balance equation models. The influence of the number of stages, temperature and residence time in each crystallizer on the particle size distribution of the product is ellucidated. Specifically, we show that for any number of MSMPRs and a constant production rate, there exists a clearly defined attainable region in a diagram of mean particle size vs. total residence time in the MSMPR cascade. This attainable region can be entirely traversed by altering the temperature and residence time in each MSMPR. We further report the influence of constraints (such as limitations in the maximally tolerated supersaturation in the MSMPRs and yield constraints) on the attainable region. By rephrasing well-established batch operating policies in the context of the attainable regoins, we highlight the underlying similarities between MSMPR cascades and batch processes and report design heuristics that can be used even in the absence of kinetic data. The concept of the attainable region and the associated heuristics could evolve into a convenient design tool for the conceptual design of continuous crystallization processes in a cost- and time-efficient manner.

Since the above mentioned investigations are dependent on the kinetics of crystal nucleation and growth, we report several case studies that employ different kinetic models and different ways to induce crystallization; among them the cooling crystallization of paracetamol from ethanol and the combined anti-solvent/cooling crystallization of aspirin from ethanol (solvent)/water (anti-solvent).

References:

[1] Randolph, A. D., Larson, M. A., Theory of Particulate Process: Analysis and Techniques of Continuous Crystallization, 2nd ed.; Academic Press: Toronto, Canada, 1988.

[2] Lindenberg, C., Krättli, M., Cornel, J., Mazzotti, M., Cryst. Growth. Des. 9, 1124-1136 (2009).

[3] Ward, J.D., Mellichamp, D.A., Doherty, M.F., AIChE J. 52, 2046-2054 (2006).

[4] Wong, S.E., Tatusko, A.P., Trout, B.L., Myerson, A.S., Cryst. Growth Des. 12, 5701-5707 (2012).

[5] Zhang, H., Quon, J., Alvarez, A.J., Evans, J., Myerson, A.S., Trout, B.L., Org. Process Res. Dev. 16, 915-924 (2012).

(241e) On the Fines Removal in a Continuous Plug Flow Crystallizer: Optimal Spatial Temperature Profile With Dissolution

Majumder, A., Loughborough University
Nagy, Z. K., Purdue University
Originally presented on: 11/5/2013 9:58:00 - 10:20:00

Fine crystals from a crystallizer can cause reduction of the efficiency of the downstream processes as well as final product quality. In this work, a systematic study is performed for obtaining the optimal temperature profile for the reduction of fines in a continuous plug flow crystallizer (PFC). The PFC consists of multiple modules or segments where the temperature of the each segment can be controlled separately. An optimization problem is formulated for a target crystal size distribution (CSD) (without fines) with the temperature of the PFC segments as decision variables. It is found that for the crystallization kinetics considered here, the optimal temperature profile also introduces dissolution steps so that the crystal fines due to secondary nucleation can be reduced significantly. A systematic study on the growth and dissolution kinetics is also performed. This study suggests that the key factor that determines whether the dissolution steps will be successful in reducing fines (without compromising the final size of the crystals from seed) is the size dependence of the growth and dissolution kinetics. If kinetics of the system is such that the larger crystals grow faster than the smaller ones and the smaller crystals dissolve faster than the larger ones, best results for fines removal is achieved. On the other hand, when both the growth and dissolution kinetics are independent of crystal size, fines removal found to be ineffective by temperature cycling.

(241f) Design and Operation of Crystallization-Based Enantioselective Processes: Enantioselective Crystallization of Chiral Compounds On An Enzymatic Reactive Environment

Encarnacion-Gomez, L. G., Georgia Institute of Technology
Bommarius, A. S., Georgia Institute of Technology
Rousseau, R. W., Georgia Institute of Technology
Originally presented on: 11/5/2013 10:20:00 - 10:42:00

Crystallization is a widely used separation and purification technique for a vast number of specialty chemicals and pharmaceutical compounds. However, its application on enantioselective separations can become troublesome. Enantiomeric compounds have the same physiochemical properties (other than optical rotation); as a result, an excess of the desired enantiomer is necessary to obtain enantiopure crystals.

To achieve the necessary enantiomeric excess, we are performing enantioselective enzymatic reactions on supersaturated solutions. To increase the productivity of the process, reaction and separation are performed in parallel in a single vessel. Preliminary work has shown the possibility of performing such a process utilizing chemoenzymatic reactions and kinetic resolutions. Present work is focused on using reaction and crystallization kinetics to design processes in which the operational time is minimized. As a case study, we are investigating the enantioselective deacylation of N-acetyl-DL-methionine by L-aminoacylase. The solubility of the DL-methionine has been measured as a function of substrate concentration, enantiomeric excess, and temperature. The information is being used to set the boundaries of the operation in which an enantiopure crystal can be recovered. The concentration in the reactive crystallizer is analyzed online by combining optical rotation, refractive index, and focused beam refractive measurements (FBRM). These signals can be used to monitor solution and crystals properties. Hence, we can control the chemical and enantiomeric purity of the crystals and their respective size distribution.

(241g) From Batch to Continuous Crystallisation: Nucleation and Growth Under Various Flow Conditions

Schacht, U., University of Strathclyde
Briggs, N., University of Strathclyde
Florence, A. J., University of Strathclyde
Sefcik, J., EPSRC Centre for Innovative Manufacturing in Continuous Manufacturing and Crystallisation University of Strathclyde
Originally presented on: 11/5/2013 10:42:00 - 11:04:00

Crystallisation processes are widely used for separation and purification in chemical and pharmaceutical industries. In order to develop continuous crystallisation of desired polymorphic forms, it is crucial to better understand the influence of mixing and flow conditions on nucleation of various polymorphs from supersaturated solution. For example, both supersaturation and post-mixing flow conditions can be used to control competitive nucleation of polymorphic forms in reactive crystallisation of L-Glutamic acid at constant temperature. We investigated various flow units for treatment of supersaturated L-Glutamic acid solutions prepared under identical initial mixing conditions in either reactive or antisolvent crystallisation.

We obtained respective crystal forms of L-Glutamic acid (either metastable α or stable β) by exposing supersaturated solutions to various flow units, such as a stirred tank, a magnetically stirred crystalliser, a peristaltic pump recycle loop or an oscillatory baffled crystalliser, operating at constant temperature. Using suitable flow units as nucleators to produce seeding suspensions of desired polymorphs, we designed a continuous crystallisation process by connecting a mixing unit (such as a static or in-line mixer) with a nucleator unit (a flow unit) and a continuous crystal growth unit (based on cooling crystallisation in a continuous oscillatory baffled crystalliser).

This set up was successfully used for continuous crystallisation of the stable β polymorph of L-Glutamic acid, achieving a good control over the final crystal size with a narrow particle size distribution.

(267b) Career Development: It Ain't A Random Walk

Worstell, J. H., Worstell and Worstell, Consultants

Most chemical engineers do not realize the difference between a job and a career. A job is a forty year random walk through the chemical processing industry. A career is a directed path to an ultimate goal. There is a second feature about careers that chemical engineers do not recognize: careers begin with a written plan. But, to write a career development plan, the chemical engineer must understand corporate structure. This presentation outlines corporate structure in terms of leadership; i.e., direct leadership, organization leadership, and strategic leadership. It then shows how to use this structure to formulate, develop, and manage a career in the chemical processing industry. This presentation also discusses the content of such a career plan and the need for it to contain three levels based on a chemical engineer's twenty year goals, which dictates the ten year goals that in turn dictate the five year goals. This presentation also discusses the need for such a plan to be evergreen.

(267a) An Industry Career in Process Development: Leveraging Chemical Engineering Training

Powell, J. B., Shell Projects & Technology

Process development is a key skill for industry, bridging from innovation to commercialization. Skills required span the universe of chemical engineering. The broad exposure provides a path to lifelong learning, and upward mobility in the organization. This poster will present an overview of process development as a career path, and discuss key success factors for managing a process development career.

(267d) Chance Favors The Prepared Mind

Hill, M., Columbia University

The author's chemical engineering career has followed a tortuous path, including episodes as a college dropout, a research engineer for a soap company, a strategic leader in the consumer products industry, and a lecturer at an Ivy League university. He will share some anecdotes from his career that illustrate important lessons for today's young professionals.

(269a) A Product Engineering Approach: Diffusion and Performance of Fragranced Products

Teixeira, M. A.,
Rodrigues, A. E., LSRE - Laboratory of Separation and Reaction Engineering - Faculty of Engineering - University of Porto

Originally presented on: 11/5/2013 8:30:00 - 8:55:00

Product Engineering Approach for the Design of High Performance Fragranced Products

Miguel A. Teixeira and Alirio E. Rodrigues*

LSRE -- Laboratory of Separation and Reaction Engineering, Associate Laboratory, Dept. of Chemical Engineering, Faculty of Engineering of University of Porto

Rua Dr. Roberto Frias, 4200-465 Porto, Portugal

Abstract

Chemical Product Engineering has emerged over the last decade as a new paradigm on Chemical Engineering.[1, 2] In fact, this is an answer to a global change in the chemical industry economy: in the past driven by technological expansion for the large production of commodities, towards a market-driven production of high added-value, specialty and performance products as happens today.[2, 3] The relevance of introducing perception, and namely consumer perception, in the design of chemical products has been recently highlighted as an important step for product design. Among the different types of perception, those related to chemical perception (chemosensory) and namely taste and olfaction are still difficult to work with. This is due to the complexity of the process of olfaction at the biochemical level.

Fragrances are presented in countless products we deal with everyday like perfumes, shampoos and soaps among many others. In fact, more than three-quarters of toiletries and household products contain a fragrance in its composition. However, the design of fragranced products is still a combination of art, technology, and some scientific knowledge, although the former predominates at the industry.[4, 5]

In order to reduce subjectivity and empiricism in this field, we have developed an integrated theoretical model to assess the performance of fragrances considering the release and diffusion processes together with the predicted odor intensity over time and distance from the source. In order to achieve that goal, our model combines different scientific fields: Thermodynamics, Transport Phenomena, and Psychophysics into a common product - fragrances. This approach is within Product Design or Engineering, which in this case is applied to Perfume Engineering as shown in Figure 1.

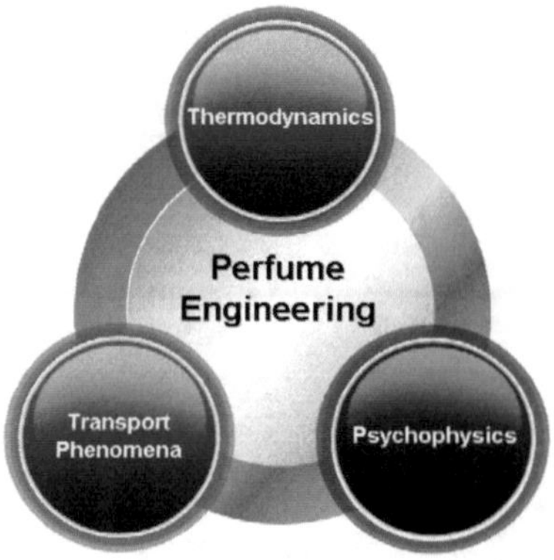

Figure 1 - Our vision of what is Perfume Engineering: the integration of concepts from different fields (Thermodynamics, Transport Phenomena, and Psychophysics) in order to improve the design and development of new perfumed products.[4]

Within this line of though, we can simplify fragranced products to a mixture of *N* fragrance raw materials diluted in a suitable matrix (a liquid solvent, a water/oil emulsion among others) at some concentration which can be sprayed on the skin or the clothes by the customers. After application of the product, the process of perception proceeds as follows: *i)* fragrances start to evaporate at different rates depending on their physicochemical properties and those of the mixture and, then, *ii)* they diffuse through the surrounding air until reaching the nostrils of consumers or other people around; and, finally, *iii)* fragrance molecules will eventually reach the human nose and above some concentration (known as the odor threshold) they will be detected and recognized with a specific intensity and character. Thus, our approach for Perfume Engineering combines knowledge from Chemical Engineering (Thermodynamics and Transport Phenomena) which can describe steps *i)* and *ii)*, together with models from Psychophysics for step *iii)* as shown in Figure 2.[6-9]

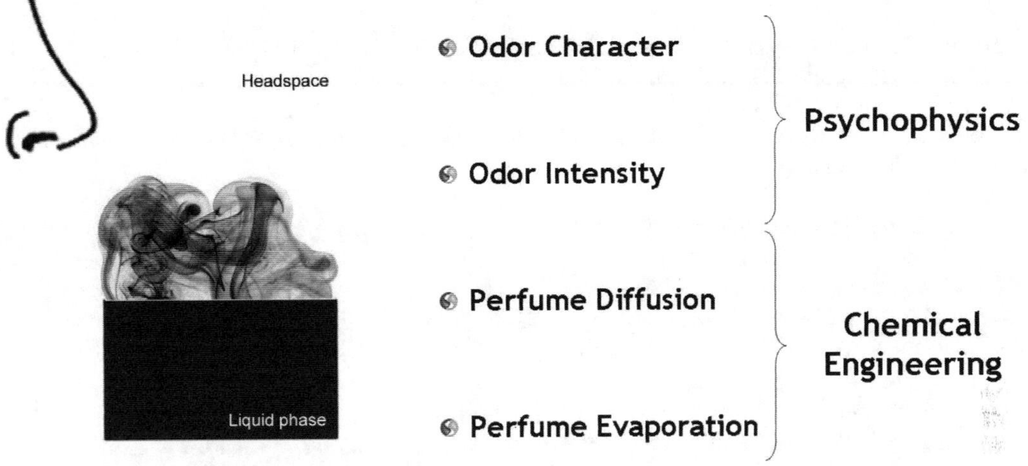

Figure 2 -- Schematic representation of our odor model for the perception of fragranced products.

Overall, our model will allow predicting and mapping the perceived odor elicited from any mixture of fragrance ingredients, thus reducing significantly the number of trial and error experiments which are expensive and time consuming for the Flavor & Fragrance industry.

In this work, we present a theoretical approach and experimental validation of our model for the step of fragrance diffusion together with the assessment of product performance using similar properties to assess it as the industry (like impact, tenacity, diffusion and volume parameters).[9] In this way, concentration profiles were experimentally measured in a diffusion tube, similar to the Stefan tube, and predicted with a model for pure fragrance chemicals, binary, quaternary, and multi-component (up to 11 chemicals) mixtures. A very good agreement between our purely predictive model and experimental concentration data was observed. Fragrance concentrations in air were then converted into odor intensities using models from psychophysics, making it possible to evaluate the evolution of the odor with time and distance using a performance plot. Accordingly, the performance of these mixtures was modeled and experimentally validated, which constitutes a landmark for fragrance design.

It was observed that the performance of a fragrance depends on the intrinsic properties of its constituents, but also on their molecular interactions in the liquid solution. Thus, the evaporation rate of each fragrance remains a function of mixture's initial composition (x_i), molecular interactions (activity coefficient, Ã¯Â □μ_i), and temperature (*T*) but it is simultaneously dependent on the physicochemical and psychological properties of each fragrance chemical (diffusivity, saturated vapor pressure, molecular weight, ODT, and power law exponent). Another important conclusion to draw from this work is that it is possible to measure and predict the performance of a perfumed product using perfumery terminology and a theoretical model to assess the perceived odor over time and distance from the source. Finally, it is expected that this work might contribute to reduce the time needed for the design of perfumes and fragranced products in the pre-formulation step and the reduction of the consumption of raw materials. Altogether, that will contribute to decrease production costs.

References

1. Cussler, E. C.; Moggridge, G. D., *Chemical Product Design*. Cambridge University Press: Cambridge, 2001.

2. Wesselingh, J. A.; Kiil, S.; Vigild, M. E., *Design & development of biological, chemical, food and pharmaceutical products*. Wiley: Chichester, U.K., 2007.

3. Cussler, E. L.; Wagner, A.; Marchal-Heussler, L., Designing Chemical Products Requires More Knowledge of Perception. *AIChE Journal* **2010,** 56, (2), 283-288.

4. Teixeira, M. A.; Rodriguez, O.; Gomes, P.; Mata, V.; Rodrigues, A., *Perfume Engineering: Design, Performance & Classification*. Elsevier: Oxford, UK, 2013.

5. Sell, C. S., On the unpredictability of odor. *Angewandte Chemie-International Edition* **2006,** 45, (38), 6254-6261.

6. Teixeira, M. A.; Rodriguez, O.; Mata, V. G.; Rodrigues, A. E., The Diffusion of Perfume Mixtures and Odor Performance. *Chemical Engineering Science* **2009,** 64, 2570-2589.

7. Teixeira, M. A.; Rodriguez, O.; Mata, V. G.; Rodrigues, A. E., Perfumery Quaternary Diagrams for Engineering Perfumes. *AIChE Journal* **2009,** 55, (8), 2171-2185.

8. Teixeira, M. A.; Rodriguez, O.; Mota, F. L.; Macedo, E. A.; Rodrigues, A. E., Evaluation of group-contribution methods to predict VLE and odor intensity of fragrances. *Industrial & Engineering Chemistry Research* **2011,** 50, 9390--9402.

9. Teixeira, M. A.; Rodriguez, O.; Rodrigues, A. E., Diffusion and Performance of Fragranced Products: Prediction and Validation. *AIChE Journal* **2013,** (accepted).

(269b) Qsrr Application to Two-Dimensional Gas Chromatography

Kamma, K., The University of Tokyo
Kaneko, H., The University of Tokyo
Funatsu, K., The University of Tokyo

Originally presented on: 11/5/2013 8:55:00 - 9:20:00

QSRR application to two-dimensional gas chromatography

1. Introduction

Gas chromatography is a widely used identification or elucidation method. In gas chromatography, a target sample is carried through a column filled with stationary phase, then retention time of compounds in the sample is observed. The structure of each compound in the sample can be identified based on its retention time, if the retention time of candidate compounds which are considered to be included in the sample is known from past observation data or from advance measurements. However, in this way, it is impossible to identify the structures of the compounds whose retention time is unknown.

Hence quantitative structure-retention relationship(QSRR)[1] is proposed to estimate the chromatographic retention time of any compounds. QSRR is a regression method of modeling the chemical structures and the retention time. First in structure descriptor calculation, chemical structures are converted into values which can be dealt with mathematically. Then a regression model is constructed between the retention time as an objective variable and the molecular structure descriptors as explanatory variables. In prediction, a new chemical structure is converted into structure descriptors and the constructed model predicts its retention time from the structure descriptors. A. Morsali et al. suggested QSRR models for BTEX and substituted benzenes based on chemical parameters.[2] Y. Zhang proposed QSRR models for polybrominated diphenyl ethers.[3]

Though QSRR is a useful method for the structure elucidation, large errors of the predicted retention time can be one of the crucial problems. Therefore it is important to construct a highly accurate model. But for the accurate models, prediction errors still remain to some degree. To deal with this problem, QSRR for two-dimensional chromatography[4] is considered useful. In two-dimensional chromatography, two different types of columns connected in line are used for retention time observation. The total retention time through the two columns is observed at the exit of second column, then retention time in each column is obtained. Constructing QSRR models for the retention time in each column enables the structure elucidation based on two predicted retention time. If one of the predicted retention time has large errors, the structure elucidation is still possible based on another predicted retention time.

The objective in this research is to improve the structure elucidation ability of two-dimensional gas chromatography by constructing QSRR models with high accuracy. We applied the partial least squares (PLS) method, the genetic algorithm-based PLS (GAPLS) method, ensemble PLS (EPLS) method to the data observed by two-dimensional gas chromatography. The prediction accuracy of the PLS, GAPLS and EPLS models is compared.

2. Method

2.1. PLS

PLS is a linear regression method. PLS extracts components which have strong correlation with objective variables from explanatory variables. Then the relationship between extracted components and objective variables is modeled. PLS can construct a robust model with high accuracy.

2.2. GAPLS

GA is an optimization method which imitates the evolution process. GAPLS is a variable selection method combining GA and PLS. GAPLS can efficiently optimize the combination of explanatory variables so that the prediction accuracy of the model improves.

2.3. EPLS

In EPLS, some variables are selected randomly from explanatory variables and many submodels are constructed between the selected variables and objective variables with the PLS method. Each submodel predicts values of objective variables, and mean or median of the predicted values are the final predicted values. The EPLS method can decrease the affection of prediction errors and increase the prediction accuracy of the model.

3. Result

3.1. Data

The set of retention time of 155 compounds was observed by two-dimensional gas chromatography. A non-polar column and a mid-polar one were used as the first column and the second column, respectively. The retention time in the first and second columns was set to be lower than 60 minutes and 8 seconds, respectively.

A software Dragon6[5] was used for structure descriptor calculation. The descriptors which failed to calculate were deleted. The descriptors in which more than 50% samples had the same value were deleted. Correlation coefficients between each descriptor are calculated, and one of the two descriptors whose correlation coefficient was over 0.9 was deleted. In the end, 164 descriptors were left.

100 compounds selected randomly were used as the training data, and the left 55 compounds were used as the test data.

3.2. QSRR Model Construction

The PLS, GAPLS and EPLS models were constructed with the training data and the models predicted the retention time of the test data. Table 1 shows the results. The retention time in the first and second column is represented as Y1 and Y2, respectively. R^2 is the evaluation criterion of the accuracy of the models. Q^2 is the cross-validated R^2 calculated by 5-fold cross-validation and R^2_{pred} is R^2 for the test data. When these values are close to 1, the model is accurate in prediction.

In prediction on Y1, R^2, Q^2 and R^2_{pred} values of PLS model were 0.984, 0.941 and 0.957, respectively. This result means that the prediction accuracy of PLS model for Y1 is high. The GAPLS model and the EPLS model also had high prediction accuracy for Y1.

In prediction on Y2, R^2, Q^2 and R^2_{pred} values of PLS model were 0.925, 0.743 and 0.676, respectively. These values mean that the prediction accuracy of PLS model is not very high. It is considered that some unnecessary variables in the explanatory variables decreased the prediction accuracy. For the GAPLS model, R^2, Q^2 and R^2_{pred} values were 0.929, 0.882 and 0.754, respectively. R^2_{pred} value of the EPLS model was 0.740, which was close to that of the GAPLS model. It can be said that the GAPLS and EPLS methods obviously improved the prediction accuracy.

In any models, the prediction accuracy for Y2 was lower than that for Y1. As a reason for that, it is considered that the observation errors of Y2 decreased the prediction accuracy. Because the second column was short so that the measurements took less than 8 seconds, observation errors of Y2 could be large.

Table 1. Modeling and prediction result

	Y1			Y2		
	R^2	Q^2	R^2_{pred}	R^2	Q^2	R^2_{pred}
PLS	0.984	0.941	0.957	0.925	0.743	0.676
GAPLS	0.990	0.983	0.966	0.929	0.882	0.754
EPLS	-	-	0.953	-	-	0.740

4. Conclusion

We aimed to construct highly accurate QSRR models for two-dimensional gas chromatography by using the PLS, GAPLS and EPLS methods. As a result, we could construct the Y1 and Y2 prediction models having high performance although the prediction accuracy of Y2 was lower than that of Y1.

In this research, we assumed that the candidate compounds included in the target sample are already presumed. But mostly, it is difficult to presume the candidate compounds in advance. In the case that candidate compounds are totally unknown, inverse analysis of QSRR models can be applied to predict candidate chemical structure based on retention time and the constructed QSRR models. It is expected that inverse analysis achieves more useful structure elucidation.

5. Acknowledgement

The authors acknowledge the support of Asahi Kasei Corp. for offering the data of two-dimensional gas chromatography.

6. References

[1] R. Kaliszan. Structure and Retention in Chromatography A Chemometrics Approach. CRC Press. May 22, 1997

[2] A. Morsali, S. A. Beyramabadi, M. R. Bozorgmehr, M. Raanaee, B. Keyvani and G. R. Jafari. Prediction of gas chromatography retention of BTEX and other substituted benzenes based on quantum chemical parameters. Scientific Research and Essays Vol. 5 (3), pp. 349-351, 4 February 2010

[3] Ya-Hui Zhang, Shu-Shen Liu, Hong-Yan Liu. Predicting the Gas Chromatographic Relative Retention Time of Polybrominated Diphenyl Ethers by MEDV-13 Descriptors. Chromatographia Vol. 65, Issue 5-6, pp 319-324, March 2007

[4] T. Hyötyläinen. Comprehensive Two-Dimensional Chromatography. SPRINGER, BERLIN. February 2013

[5] Talete srl. http://www.talete.mi.it/ 13 May 2013

(269c) Method of Moments for Computational Microemulsion Analysis and Prediction in Tertiary Oil Recovery

Fraaije, J. G. E. M., Culgi BV
Tandon, K.,
Jain, S.,
Handgraaf, J. W., University of Trieste
Buijse, M.,
Originally presented on: 11/5/2013 9:20:00 - 9:45:00

We discuss the application of Helfrich's surface torque density concept to microemulsion design and analysis, from three different angles: (i) from the point of view of coarse-grained molecular simulations, using Dissipative Particle Dynamics, including charge interactions and added salt, (ii) using an approximate double film model for the surface, and (iii) comparison with formulation approaches. The simulations use that surface torque can be calculated unambiguously from the stress profile, provided the surface is tensionless. Very good agreement is found on predicting optimal salinity (or the absence of that) for a range of surfactants: Dioctyl Sodium Sulfosuccinate (AOT), various twin tailed sulfonates and Sodium Dodecyl Sulfate (SDS). The simulations are very fast, on par with times for experiments, thus they could lead to a practical tool for discovery of more efficient surfactants, although much remains to be done with respect to other important variables: oil composition, surfactant mixtures, aggregation in solution and so on. The microscopic model (second approach) is highly approximate: it is essentially based on two opposing swelling tendencies, that are both of osmotic nature. According to the model, the tails are swollen by the oil, and the charged head groups are confined in a salty layer in Donnan equilibrium with the salt solution. In this way, the surface interactions are purely entropic. The comparison of the film model with existing formulation approaches (third approach) covers: interfacial tension minimum, Winsor R theory, Quantitative Structure Property Relations (QSPR), Hydrophilic-Lipophilic Deviation (HLD), and HLD-Net Average Curvature, and temperature coefficients. Using the surface torque analysis, we succeed in deriving in an 'ab initio' way QSPR empirical coefficients that have been known for decades, but, until now, have been obscure in origin.

(269d) A Computational Methodology for Designing Mixtures

Austin, N.,
Samudra, A., Rockwell Automation
Sahinidis, N., Carnegie Mellon University
Trahan, D. W., Massachusetts Institute of Technology
Originally presented on: 11/5/2013 9:45:00 - 10:10:00

As trial-and-error methods are becoming increasingly intractable with the growing library of chemical compounds, computer-aided molecular design (CAMD) is a necessary tool for the future of chemical production. CAMD combines optimization techniques with many property prediction models, allowing for a rapid and accurate search through all chemical structures, even those for which no data is available. These techniques work well for designing single molecules, but a more difficult –and more applicable – problem arises when attempting to design mixtures [1].

We propose a novel mixture design approach which couples molecular design with derivative-free and algebraic model-based optimization. We use derivative-free optimization (DFO) to search in the space of compound properties. Requirements on properties of the individual components are then converted into molecular structures via our software AMODEO and a recently developed molecular design methodology [2], which relies on Group Contribution methods for property estimation [3]. Finally, we use mixing rules and algebraic global optimization in order to optimize over the composition space for a given set of mixture components. The approach exploits a key insight to the molecular mixture design problem, namely that the small number of degrees of freedom in the property space can be dealt with efficiently via derivative-free optimization (DFO) algorithms [4]. Several case studies are presented to illustrate the wide applicability of the proposed methodology.

References:

1. Conte, E., R. Gani, and K. M. Ng, Design of formulated products: A systematic methodology, AIChE Journal, 57, 2431-2449, 2011
2. Samudra, A. and N. V. Sahinidis, Optimization-based framework for computer-aided molecular design, AIChE Journal, accepted, 2013
3. Marrero, J. and R. Gani. Group-contribution based estimation of pure component properties, Fluid Phase Equilibria, 183–184:183–208, 2001
4. Rios, L. M. and N. V. Sahinidis, Derivative-free optimization: A review of algorithms and comparison of software implementations, Journal of Global Optimization, DOI 10.1007/s10898-012-9951-y, 2012

(269e) Developing a QSAR Relating Solvent Structure to Crystal Morphology Using 2D & 3D Molecular Descriptors

Haser, J. C., Auburn University
Herring, R. III, Auburn University
Eden, M. R., Auburn University
Originally presented on: 11/5/2013 10:10:00 - 10:35:00

Crystallization is one of the least-studied separation techniques compared to distillation and extraction processes, especially in computer modeling and prediction. There are many more variables in determining final quality (crystal morphology, crystal purity, potential recovery) in crystallization processes than distillation and extraction [1]. Crystallization process products are typically produced in small quantities and their properties can have a large effect on the downstream processes. Therefore it is important that crystals meet strict guidelines in final quality.

Studies have been done to predict ibuprofen crystal morphologies using linear models based on hydrogen bonding properties of several different pure solvents [2]. The goal of this work is to develop a predictive model of crystal morphology based on solvent choice. The model will be developed using Principal Component Analysis and multivariable linear regression. Using a combination of 2D & 3D molecular descriptors is ideal for crystallization solvent design because the structure of the solvent is critical to the resulting morphology of the crystal [2]. With this model, the relationship between solvent molecular geometry and crystal aspect ratio can be determined. The first work will be focused on the aspect ratio of ibuprofen crystals and then further expanded to account for any solute molecule. It is advantageous to be able to determine what solvent characterization will produce crystals of the desired quality, and this method will allow solvents to be chosen without using time- and money-consuming experimental measurements.

[1] Karunanithi A.T., Achenie, L.E.K, and Gani, R. (2006) A computer-aided molecular design framework for crystallization solvent design. Chem. Eng. Sci., **61**, pp. 1247–1260.

[2] Acquah, C., Karunanithi, A.T., Cagnetta, M., Achenie, L.E.K., and Suib, S.L. (2009) Linear models for prediction of ibuprofen crystal morphology based on hydrogen bonding propensities. Fluid Phase Equilibria, **277**, pp. 73-80.

(269f) Achieving Target Emulsion Drop Size Distributions Using Population Balance Equation Models of High Pressure Homogenization

Maindarkar, S. N., University of Massachusetts
Dubbelboer, A., University of Technology
Hoogland, H., Unilever R & D
Henson, M. A., University of Massachusetts Amherst
Originally presented on: 11/5/2013 10:35:00 - 11:00:00

Oil-in-water emulsions are ubiquitous dispersed phase systems with diverse applications that include consumer products, processed foods, polishes, waxes, agricultural sprays and road surfacing materials. In the foods industry, emulsions constitute natural foods as well as numerous processed products such as milk, butter, margarine, ice cream, sauces and desserts. Food emulsions contain edible oils, water and biocompatible surfactants as the major ingredients and vitamins, minerals and/or flavors as minor ingredients. Emulsion system formulation and processing operations both impact the drop size distribution, a key property that influences emulsion rheology, stability, texture and appearance. A typical processed food requires the drop size distribution to be maintained within acceptable limits, which includes achieving a prescribed mean drop size, maintaining small variations about the mean and avoiding very small or large drops that adversely affect product properties such as texture, appearance and food safety. Oil-in-water emulsions are typically formed by first preparing a coarse premix using a low shear stator-rotor type device that mixes the various ingredients into a stable form. This premix is then processed with a high shear device in which relatively large drops are broken into much smaller drops. In high pressure homogenization, the coarse emulsion is passed through a small orifice under very high pressure.

Due to lack of quantitative understanding, new emulsified products are currently developed by combining a broad knowledge of previous product formulations with empirical scientific experimentation. An alternative to brute force experimentation is to utilize a suitable mathematical model to predict the drop size distribution for different emulsion formulations and processing conditions. The population balance equation (PBE) modeling framework is particularly well suited for this problem as functions describing drop breakage and coalescence can be incorporated within a fundamental number balance equation to predict the evolution of the drop size distribution. Several investigators have developed PBE models of high pressure homogenizers. As part of our previous work, we have developed breakage-only PBE models for prediction of the drop volume distribution. Through experimental and computational studies, we showed that our model emulsion system exhibited negligible coalescence due to the low oil-to-surfactant ratio (5% oil, 1% surfactant) used.

In this paper, we incorporate drop coalescence functions to allow the drop volume distribution to be predicted for a much higher oil-to-surfactant ratio (50% oil, 1% surfactant) and utilize the model to achieve target distributions by changing homogenizer operating conditions. Nonlinear optimization is used to estimate six adjustable parameters in the drop breakage and coalescence functions from measured size distributions. The experimentally validated model is used to optimally determine the number of homogenization passes and the pressure of each pass such that specified emulsion drop size properties are achieved in a least-squares sense. Two optimization objectives that differ with respect to the distribution specifications are formulated and experimentally evaluated.

(298a) Applying Crystallization Modeling to Improve the Understanding of a Batch Cooling Process of An Agrochemical Active Ingredient

Brent, R.,
Parmar, M., Syngenta
George, N., Syngenta
Sillers, P.,
Bryce, M.,
Clelland, G.,
Mumtaz, H., Process Systems Enterprise
Mitchell, N., Process Systems Enterprise
Bermingham, S. K., Process Systems Enterprise
Originally presented on: 11/5/2013 12:30:00 - 12:52:00

The current production process of a Syngenta active ingredient isolates the batch by a cooling crystallization process. A major bottleneck of this process is the isolation time of the product after crystallization, which can vary significantly between batches. From observation, the key variable that dictates the isolation time of the product is the width of the particle size distribution of the crystals obtained; the wider the particle size distribution, the slower the filtration time observed. Typically for this product, two distinct environments of fine particles and large single crystals are observed within a single batch. In this work an experimental scale-down study was carried out to replicate the plant process and to assess whether the filtration properties could be improved by modifying the conditions in the crystallizer. Seeding the batch with different quantities of seed with a narrow particle size and modification of the cooling profile where carried out, with limited success in narrowing the particle size distribution significantly. Crystallization modeling using gCRYSTAL was employed to rationalize the experimental observations and to identify the key crystallization phenomena that are dominant in the process. It was observed that the controlling phenomena for this crystallization where the favorable kinetics for crystal growth and also attrition of the crystals, once attaining crystals that had grown to a critical size. Having identified this from the modeling, it was then possible to devise additional experiments to reduce the width of the particle size distribution of this material with the potential to improve plant operations.

(298b) Generic Modelling Framework for Design and Analysis of Crystallization Operations

Meisler, K. T., Technical University of Denmark (DTU)
Samad, N. A. F. A., Technical University of Denmark (DTU)
Gernaey, K. V., Technical University of Denmark (DTU)
Gani, R., Technical University of Denmark
Originally presented on: 11/5/2013 12:52:00 - 13:14:00

A systematic model-based approach to design and analysis of crystallization operations as a first step could result in reduction of time and experimental resources. To this end a generic and systematic model-based framework for the design of a process monitoring and control system to achieve the desired crystal size distribution (CSD) and crystal shape for a wide range of crystallization processes has been developed. This framework combines a generic multi-dimensional modelling framework, tools for design of supersaturation set-point, for design of PAT systems as well as options to perform the uncertainty and sensitivity analysis of the PAT system design. Through this framework, it is possible for a wide range of crystallization processes, to generate the necessary problem-system specific models, the necessary supersaturation set-point using an extended analytical CSD estimator and/or a response surface method (RSM) and a PAT system design including implementation of monitoring tools and control strategies in order to produce a desired product with its corresponding target properties. The problem-system specific models involve generic elements and system specific elements. The system specific elements, such as kinetic constitutive models (models for growth, nucleation, agglomeration and breakage) are stored in libraries, from which they can be extracted, analyzed, used and re-used. This allows models to be analyzed against other models for specific systems and the possibility of comparing models for different systems in a fast and flexible manner. An advantage of this is the possibility to study the effect of different constitutive model parameters on the kinetic phenomena associated with different chemical systems together with the crystallization operational parameters and employ this knowledge to plan and design crystallization operations. Uncertainty and sensitivity analysis is also possible within the framework, and can be carried out to determine, for example, Â theÂ kinetic model parameters in the growth rate model that are the most sensitive This presentation highlights the advantages of utilizing the generic modeling framework to support crystallization operation modeling and the use thereof in design and analysis of crystallization operations. This is illustrated through simulation studies of a potassium dihydrogen phosphate (KDP) and a paracetamol crystallization operation. The operation of KDP batch cooling crystallization is simulated with varying seeding policies and temperature profiles in order to establish the effects of different operational policies on the final product. This includes the effects of uncertainties in the input, for example in the kinetic model parameters. The advantage of using a framework for which models can be re-used is highlighted as the user is guided in specifying the needed data under the assumptions stated in the modeling problem through the formulation and solution of the model equations representing the operation scenario under investigation. The advantage of the generic framework is further highlighted when the chemical system is switched to paracetamol for a similar operation scenario. Here, switching the kinetic models within the operation scenario model allows the study of paracetamol crystallization operation, in one as well as two-dimensions.

(298c) Crystallization Process Optimization Using Derivative-Free Optimization Algorithms

Zhao, H.,
Vouzis, P., Carnegie Mellon University
Sahinidis, N., Carnegie Mellon University
Originally presented on: 11/5/2013 13:14:00 - 13:36:00

Polymorphism is of key importance in the pharmaceutical and fine-chemicals industry. Modeling of polymorphism phenomena is based on Population Balance Equations (PBEs), which are sets of partial differential equations whose analytical solution is not always available [1]. Moreover, when optimization is to be done over a simulator, it is often difficult to calculate derivatives of variables of PBEs either analytically or numerically. This renders most derivative-based optimization methods of little or no use. Derivative-free optimization (DFO) algorithms are specifically designed for this kind of problem [2]. In this work, we use g-CRYSTAL as a black box simulator to perform simulation-based optimization of crystallization processes. DFO is used to optimize one particular polymorph and minimize the time of producing a desired particle size distribution. This goal is achieved by optimizing process variables, including temperature and cooling profile of reactor, super saturation of seed molecule, flow rates, and concentrations of species. We present extensive computational results, where we compare the results of 28 DFO algorithms on several case studies.

References

[1] Rudiyanto Gunawan, Irene Fusman, and Richard D. Braatz. High resolution algorithms for multidimensional population balance equations. AIChE Journal, 50(11):2738{2749, 2004.

[2] Luis Miguels Rios and Nikolaos V.Sahinidis. Derivative-free optimization: A review of algorithms and comparison of software implementations. Journal of Global Optimization, DOI 10.1007/s10898-012-9951-y, 2012.

(298d) Stable Configurations in the Horizontal Ribbon Growth Process

Oliveros, G. A., Carnegie Mellon University
Ydstie, B. E., Carnegie Mellon University
Sridhar, S., Carnegie Mellon University

Originally presented on: 11/5/2013 13:36:00 - 13:58:00

In the Horizontal Ribbon Growth (HRG) process, a thin crystal ribbon is produced and extracted continuously from a molten substrate. The advantage of this technique is its capacity to achieve large rates of production compared to conventional crystal growth processes, such as the Czochralski and Bridgman. However, despite its potential and efforts to develop the technology for more than 50 years, many technical and experimental challenges have been reported regarding the difficulty in trying to maintain a continuous and stable ribbon production.

One of the principal causes of disruption is the freezing of the ribbon to the end of the crucible. The reason for this is that there is no melt present between the ribbon and the crucible. The solid surface of the crucible offers plenty of nucleation points for the ribbon to crystallize onto as it is being pulled. The opposite problem can also arise: an excess of melt flow that causes the melt to spill over the crucible. In an ideal operation, a space between the ribbon and the lip of the crucible must be maintained to avoid undesired nucleation points, and a meniscus must be pinned between the crucible lip and the ribbon to avoid melt spill-over.

In this work, we present a theoretical analysis to assess these technical difficulties. We analyze the conditions under which a meniscus exists between the ribbon and the crucible and stays pinned to its corner, as well as the allowable range of melt levels that give rise to stable menisci. Assuming that the existence of the meniscus is defined by hydrostatics, we use a theoretical model based on the governing Young-Laplace equation to solve analytically for the shape of the meniscus. Then, we show the range of operating conditions that are physically viable in a silicon system, and construct the possible stable configurations. We show that these stable configurations depend mainly on the pulling angle, the melt level, and the type and shape of the crucible.

(298e) Using ATR-FTIR to Track the Crystallization of Multiple Solutes

Griffin, D., Georgia Institute of Technology
Grover, M., Georgia Institute of Technology
Kawajiri, Y., Georgia Institute of Technology
Rousseau, R. W., Georgia Institute of Technology
Originally presented on: 11/5/2013 13:58:00 - 14:20:00

In-situ measurements of concentration play a large role in developing and implementing control schemes for crystallizers [1, 2]. In recent years Attenuated Total Reflectance Fourier Transform Infrared (ATR-FTIR) technology has been widely used to track the concentration profiles of a single solute during crystallization processes [3-5]. However, if the parent solution is complex and contains multiple potentially crystallizing solutes (e.g. nuclear waste) multiple solutes must be tracked simultaneously. This can be a challenging task if the characteristic infrared absorbance peaks for the individual solutes are overlapping. In this work a supervised learning approach was taken to construct a correlation between the infrared absorbance spectrum and the concentrations of multiple solutes in the solution. Furthermore, to avoid over-fitting and establish robust correlations, the training set was carefully selected and projection-based regression techniques [3, 6] were applied—of these, Partial Least Squares Regression (PLSR) performed the best.

This methodology was used to monitor crystallizations of complex electrolytic solutions representative of nuclear waste stored at Hanford, WA [7-9]. In particular, four solutes (carbonate, sulfate, nitrate, and nitrite) were tracked during batch cooling crystallizations. By monitoring a number of crystallizations, operating zones for crystallizing particular salts were identified. Furthermore, in future work, multi-solute concentration monitoring will be used to facilitate a direct design approach [10] to control the crystallization of clean (non-radioactive) sodium salts from nuclear waste mixtures.

1. Nagy, Z.K. and R.D. Braatz, *Advances and New Directions in Crystallization Control.*Annual Review of Chemical and Biomolecular Engineering, Vol 3, 2012.

2. Zhou, G.X., et al., *Direct design of pharmaceutical antisolvent crystallization through concentration control.* Crystal Growth & Design, 2006. **6**(4): p. 892-898.

3. Togkalidou, T., et al., *Solute concentration prediction using chemometrics and ATR-FTIR spectroscopy.* Journal of Crystal Growth, 2001. **231**(4): p. 534-543.

4. Lewiner, F., et al., *On-line ATR FTIR measurement of supersaturation during solution crystallization processes. Calibration and applications on three solute/solvent systems.* Chemical Engineering Science, 2001. **56**(6): p. 2069-2084.

5. Dunuwila, D.D. and K.A. Berglund, *ATR FTIR spectroscopy for in situ measurement of supersaturation.* Journal of Crystal Growth, 1997. **179**(1-2).

6. Xiaobo, Z., et al., *Variables selection methods in near-infrared spectroscopy.* Analytica Chimica Acta, 2010. **667**(1–2): p. 14-32.

7. Herting, D.L., *Clean Salt Process - Final Report*, E. Management, Editor. 1996.

8. Nassif, L., et al., *Pretreatment of Hanford medium-curie wastes by fractional crystallization.* Environmental Science & Technology, 2008. **42**(13): p. 4940-4945.

9. Rajbanshi, A., B.A. Moyer, and R. Custelcean, *Sulfate Separation from Aqueous Alkaline Solutions by Selective Crystallization of Alkali Metal Coordination Capsules.* Crystal Growth & Design, 2011. **11**(7): p. 2702-2706.

10. Fujiwara, M., et al., *First-principles and direct design approaches for the control of pharmaceutical crystallization.* Journal of Process Control, 2005. **15**(5): p. 493-504.

(298f) Operation and Optimization of Seeded Batch Crystallization When Crystal Growth Rate Is Size-Dependent

Ting, W. L., National Taiwan University
Wang, H. Y.,
Ward, J. D., National Taiwan University
Originally presented on: 11/5/2013 14:20:00 - 14:42:00

In this work, the operation and optimization of seeded batch crystallization processes is investigated when crystal growth rate is size-dependent using a dimensionless model[1] and model reported in the literature for the crystallization of potassium alum.[2] Many of the results are qualitatively similar to the case when growth rate is size independent: The seed mass and average seed size affect the results in a similar way. An approximate analytical trajectory that works well for size-independent growth also works well when the growth rate is size dependent. The optimal supersaturation trajectory is qualitatively similar to the case when growth rate is size independent. Finally, we show that an approximate, size-independent model can be used to determine the optimal supersaturation trajectory with little error. This is advantageous because optimizing the size-independent model requires much less computational effort.

References

1. Ward JD, Yu CC, Doherty MF. A New Framework and a Simpler Method for the Development of Batch Crystallization Recipes. *Aiche J.* 2011;57(3):606-617.
2. Tai CY, Lu EL, Yu KH. Determination of Secondary Nucleation Kinetics Using a Batch Crystallizer. *J. Chin. I. Ch. E.* 1991;22(4):231-239.

(298g) Effects of Several Heat Shield On the Crystallization Front of Sapphire Crystal Grown By Kyropoulos Method

Song, D. W., LG Siltron
Kim, H., University of Seoul
Ageyeva, S., CDT Equipment LLC
Originally presented on: 11/5/2013 14:42:00 - 15:04:00

In the growth of sapphire crystals, the critical density of dislocations determined by variations of the ratio of G/V affects on the quality of crystal structure. Where G is the axial temperature gradient and V is the velocity of crystallization front (CF). In general, it is not the initial temperature gradient is so important for structure control but rather the degree of its constancy maintaining during growth. However in Kyropoulos (KY) method the CF, has formed the inverse shape of cone, is gradually down to crucible bottom. Therefore it is more difficult to control its consistency during growth in KY method than other methods.

In our experiments, we have studied the influences of heat shield (HS) on the formation of low angle block or grain boundaries (LAGB) and micro-bubbles. We have estimated the angle and velocity of convex of CF in several HS configurations. The shapes of CF are determined by the patterns of micro-bubbles generated by the power down-and-up test (PDU) in quick time. And the defects such as LAGB and micro-bubbles are inspected in the vertical slug which had been cut in center line of grown crystals by polarized microscope. The velocities are measured in PDU-tests by the evaluation data of vertical slugs and calculated by SK Forecast (developed by Alyuda.com) and our simulators.

The defects free shape of CF is depended on the HS configurations with respect to time. It means that there exist an optimal tendency of CF in each HS. The LAGB is related to the acceleration of growth rate and the micro-bubble is related to the oscillation of growth rate that will be discussed in detail. The difference between our simulators and SK Forecast are come from the consideration of energy conservation and initial guessing value of angle and velocity of CF.

(308a) Design of Heat Transfer Media Components for Retail Food Refrigeration

Sahinidis, N., Carnegie Mellon University
Samudra, A., Rockwell Automation
Originally presented on: 11/5/2013 12:30:00 - 12:55:00

High levels of emissions of ozone-depleting substances and greenhouse gases from supermarkets around the world can be attributed to leakage of hydroflurocarbons (HFCs) and generation of electric power required for retail food refrigeration. Indirect refrigeration loops are ideally suited for reductions mandated by regulation standards as they reduce leakage and can lead to significantly lower total energy consumption. Hence, the design and identification
of fluids that boost refrigeration performance while meeting safety and environmental guidelines is of considerable interest. Using a recently developed molecular design methodology as our starting point, in this work, we develop a model and search technique for identifying ideal secondary refrigerants. Accurate property models that predict characteristic refrigerant properties guide the search for molecules. We also include environmental and safety metrics (biodegradability and lethal concentration (LC50)), along with performance criteria for heat transfer efficiency to analyze the candidate molecules. We identify a number of novel molecules as well as known compounds that have not been used as secondary refrigerants.

(308b) Manufacturing and Design of Laser Fields for Engineering Molecular Dynamics

Koswara, A.,

Originally presented on: 11/5/2013 12:55:00 - 13:20:00

Advances in quantum control theory and optimization as well as in ultrafast optics have enabled the dynamic control of chemical reactions using first-principles quantum chemical models. Here, shaped laser pulses, rather than chemical catalysts, are designed and manufactured to selectively break or rearrange a molecule's bonds. For the purpose of spectroscopy, laser fields may be crafted and used as photonic reagents to optically and dynamically discriminate (ODD) compounds with very similar chemical properties. Specifically, in ODD, the subtle differences between different chemicals' intramolecular structure are magnified through their unique quantum dynamical interactions with optimally designed fields. These two phenomena are examples of a new paradigm in chemical engineering design where photons rather than molecules are the products. In our group, we investigate different control schemes for the dynamical manipulation of molecular reaction and spectroscopy in gases, taking advantage of the high duty cycle of photonic reagent manufacturing. In this talk we provide an overview of the properties of photonics reagents, how they are designed, and how they are manufactured.

(308c) Multi Scale Product Design From Enzyme Design to Bio Network Control

Marimuthu, K., Carnegie Mellon University
Chakrabarti, R., Carnegie Mellon University
Originally presented on: 11/5/2013 13:20:00 - 13:45:00

An efficient product design involves modeling and optimization across length and time scales. This multi scale product design has become widespread in biological systems. In this talk we present a case study of our group's work in multi scale design of biochemical reaction networks. DNA amplification or the Polymerase Chain Reactions (PCR) is the workhorse of nearly every modern molecular biology laboratory, as well as the burgeoning discipline of personalized medicine. Though PCR is simple, this method is often fraught with difficulties that can decrease the cycle efficiency or result in competitive amplification of undesired side products. To date, nearly all approaches to PCR optimization have been based on either heuristic experimental sampling of reaction conditions or calculation of the effects of changes in reaction mixture composition on thermodynamic parameters. By contrast, the engineering discipline of control theory can automatically derive prescriptions for the optimal temperature cycling protocols of a PCR reaction, if a suitable kinetic model exists. We have developed the first sequence and temperature dependent kinetic model for PCR reactions, validating this model through comparison to experimental data. Using this kinetic model we have analyzed the controllability of PCR and developed a control strategy to maximize the DNA amplification efficiency. Further, we have formulated and solved a large scale constrained dynamic optimization problem and derive the optimal temperature profile that maximizes the amplification of a given target DNA sequence and reduces the overall reaction time. This bionetwork control problem motivates the formulation of DNA oligonucleotide design and polymerase enzyme design objectives. By solving these molecular design problems, we can further improve the amplification selectivity and efficiency. The output of the multi scale product design methodology is a temperature cycling protocol and reaction mixture composition that is tailored for the amplification of any desired DNA target.

(341a) Molecular Design of Thermomorphic Solvents for the Hydroformylation of Dodecene

McBride, K., Max Planck Institute for Dynamics of Complex Technical Systems
Sundmacher, K., Max Planck Institute for Dynamics of Complex Technical Systems
Originally presented on: 11/5/2013 15:15:00 - 15:36:00

Molecular design of thermomorphic solvents for the hydroformylation of dodecene

Kevin McBride[1], Kai Sundmacher[1,2]

[1]Max Planck Institute for Dynamics of Complex Technical Systems, Magdeburg/DE

[2]Process Systems Engineering, University of Magdeburg, Magdeburg/DE

The use of thermomorphic multi-component solvent (TMS) systems is an innovative concept to incorporate homogeneous reactions with efficient catalyst separation via a temperature induced phase split. In the rhodium catalyzed hydroformylation of long chain olefins, the TMS is a promising prospect due to the requirement that the expensive catalyst complex be recycled quantitatively. Thus, the key characteristic of the TMS is its temperature sensitivity, meaning that the phase behavior of these solvents can be tuned using temperature as the control variable. We keep the temperature high during the reaction, forming one phase, and then, with a reduction in temperature, form two liquid phases where the catalyst is selectively removed from the other components. Sought after properties are a relatively wide miscibility gap and low distribution of the catalyst. In this regard, selection of the solvent is a critical aspect of the successful implementation of such a reaction.

Current methods used to select candidate solvents for TMS do not contain molecular information. The most current approach uses Hansen parameters and temperature limitations; however, they do not contain information about the liquid-liquid phase behavior. Hence, the design space of the existing methods is limited. In addition, the effect that the product has on the miscibility gap is not considered in the preliminary evaluation. The effect of the TMS process-wide has also not been investigated.

In this work, we take both of these considerations into account. Using other models such as modified UNIFAC or COSMO-RS, which contain information about the solvents at the molecular level, solvent screening is conducted. Using a geometric analysis of the liquid-liquid equilibrium at various temperatures, candidate solvents are identified. Upon further inspection for feasibility, reactivity of the solvents, and influence of the product on the miscibility gap, screened candidates are inserted into an optimizer that finds the minimum cost of the selected process flowsheet. Here, we use the hydroformylation of dodecene as an example reaction with the goal of finding the optimal polar solvent for the TMS with decane as the non-polar component.

By using this procedure, we show that a rigorous TMS screening method combined with process-wide optimization, allows us to consider the overall effects of the TMS on cost during the initial process design.

(341b) Advanced Supported Ionic Liquid Phase (SILP) Materials for Intensified Catalytic Processes

Wasserscheid, P., Friedrich-Alexander University Erlangen-Nuremberg
Haumann, M., Friedrich-Alexander University Erlangen-Nuremberg
Originally presented on: 11/5/2013 15:36:00 - 15:57:00

Novel Supported Ionic Liquid Phase (SILP) materials consist of an ionic liquid, dispersed as a thin film on the inner surface of a highly porous solid material.[1] By dissolving homogeneous transition metal complexes in the ionic liquid film, the SILP concept allows tailor making of solid materials with definite properties and a controlled chemical reactivity. Since the ionic liquid is dispersed on the inner surface of the support, a dry solid material is obtained. These materials can be handled like classical heterogeneous catalysts and are highly attractive for large scale applications.[2]

Due to the extremely low vapour pressure of ionic liquids, the SILP concept is especially suited for continuous gas-phase reactions. No leaching of ionic liquid and catalyst can occur via the gas-phase and the SILP catalyst remains intact under steady state conditions for more than 1000 hours time on stream. Since the catalyst is retained inside the reactor, only products and non-converted substrates leave the reactor, thus simplifying the downstream processing significantly.[3]

In this contribution we highlight the latest developments from our research with SILP catalyst materials in syngas applications, namely the hydroformylation of technical feedstock containing mixed alkenes and the low temperature water gas shift reaction. Excellent performance with regard to activity, selectivity and stability is obtained, thus making these SILP catalysts highly attractive for industrial applications.

[1] A. Riisager, R. Fehrmann, S. Flicker, R. van Hal, M. Haumann, P. Wasserscheid, *Angew. Chem. Int. Ed.* **2005**, *44*, 185 (VIP).

[2] A. Riisager, R. Fehrmann, M. Haumann, P. Wasserscheid, *Eur. J. Inorg. Chem.* **2006**, 695.

[3] M. Jakuttis, A. Schönweiz, S. Werner, R. Franke, K.-D. Wiese, M. Haumann, P. Wasserscheid, *Angew. Chem. Int. Ed.* **2011**, *50*, 4492.

(341c) Development of a Novel Process and Physico-Chemical Data for the Production of Crotonaldehyde From Aqueous Acetaldehyde Solutions

Scheithauer, A., University of Kaiserslautern
Grutzner, T., Lonza AG
Rijksen, C., Lonza AG
Zollinger, D.,
Hasse, H., University of Kaiserslautern
Originally presented on: 11/5/2013 15:57:00 - 16:18:00

Crotonaldehyde is an important intermediate in several value added chains in the chemical industry, e.g. in the production of sorbic acid. The state-of-the-art process for the production of Crotonaldehyde is described in [1]. The starting material is Acetaldehyde in aqueous solution. It is converted in a two-step process, first under basic and then under acidic conditions, to Crotonaldehyde following an Aldol addition-condensation mechanism. Several intermediates occur, with Acetaldol being described as the most important. There are also side reactions. The information which is available in the literature is not sufficient to set up a reliable model of the process or to study alternatives.

Therefore, in the present work, in a first step, fundamental studies on the reaction system involved in the Crotonaldehyde production from Acetaldehyde were carried out using NMR spectroscopy. Surprisingly, it was found that Acetaldol is only present in small quantities. The reaction mechanism was elucidated and a quantitative model of the reaction network was developed. The proposed model also takes into account that Acetaldehyde reacts with water and forms oligomers. Despite the industrial importance of aqueous solutions of Acetaldehyde, the information available in literature on the chemistry in that system is poor. That gap was closed by further systematic NMR spectroscopic studies, and a new physico-chemical model for describing the vapor-liquid equilibrium of aqueous solutions of Acetaldehyde was developed based on the experimental results.

The new models were used for developing a novel highly integrated process for the production of Crotonaldehyde from aqueous solutions of Actetaldehyde which is advantageous in comparison with the established process described in [1].

[1] R. Schulz et al.: Crotonaldehyde and Crotonic Acid, *in*: *Ullmann's Encyclopedia of Industrial Chemistry, Wiley-VCH, Weinheim,* 2005, 1-8

(341d) Horizontal Reactive Distillation Concepts and Their Application for Continuous Multicomponent Chiral Resolution

Au-Yeung, P. H., The Dow Chemical Company
Resnick, S. M., The Dow Chemical Company
Witt, P. M., The Dow Chemical Company
Frank, T. C., The Dow Chemical Company
Donate, F. A., The Dow Chemical Company
Robbins, L. A., The Dow Chemical Company
Originally presented on: 11/5/2013 16:18:00 - 16:39:00

This presentation will describe a new reactive distillation technology involving a novel horizontal reactive distillation apparatus and a new overall process scheme.* The new technology enables simultaneous production and isolation of up to four enantiomers via reversible enzymatic enantioselective acylation. Reactive distillation is used to drive the reaction forward and auxiliary distillation columns are used to isolate the enantiomers. In the (*R*)-enantioselective acylation reaction scheme, both the organic substrate and the acyl donor are chiral compounds (supplied as racemates), so the process can produce two (*R*)-enantiomer reaction products plus two unreacted (*S*)-enantiomers. The effective overall reaction is:

(*R*,*S*)-A + (*R*,*S*)-B <----> (*R*)-C + (*R*)-D + (*S*)-A + (*S*)-B

The results of kinetic studies, miniplant experiments, and process simulations will be presented for a model reaction system: lipase-catalyzed (*R*)-enantioselective transesterification of (*R*,*S*)-1-*n*-butoxy-2-propanol with (*R*,*S*)-1-methoxy-2-acetoxypropane to yield (*R*)-1-*n*-butoxy-2-acetoxypropane and (*R*)-1-methoxy-2-propanol plus unreacted (*S*)-1-*n*-butoxy-2-propanol and (*S*)-1-methoxy-2-acetoxypropane. A horizontal compartmentalized reactive distillation vessel is specified instead of a conventional reactive distillation column to provide long liquid-phase residence time needed for good conversion (on the order of hours). Low vapor-traffic pressure drop within the headspace of the vessel allows operation under vacuum at reduced temperatures (about 5 mm Hg and 25 °C) for good enzyme stability and enantioselectivity. The technology has potential to enable a number of enzymatic equilibrium-limited enantioselective reactions including acylation, transamidation, and transamination – as a means to producing a variety of small molecule chiral compounds including hydroxy compounds, esters, amines, and amides. In principle, these compounds may be used as chiral synthons (synthetic building blocks) in the asymmetric synthesis of biologically-active products such as pharmaceuticals, agricultural chemicals, and flavor and aroma chemicals. More broadly, this new approach to conducting reactive distillation may be useful for any system requiring long liquid-phase residence time or low vapor-traffic pressure drop.

* Additional background is available in an article accepted for publication in *AIChE Journal* (2013). DOI: 10.1002/aic.14138. Sol Resnick is now with Calysta Energy, Menlo Park, California, U.S.A.

(341e) Reactive Distillation With Biocatalytic Coatings for Structured Packings

Heils, R., Hamburg University of Technology
Hilterhaus, L., Hamburg University of Technology
Liese, A., Hamburg University of Technology
Smirnova, I., Hamburg University of Technology
Originally presented on: 11/5/2013 16:39:00 - 17:00:00

Introduction

Reactive distillation is an integrative separation process that is well-established in industry for diverse etherification and esterification reactions (e.g. MTBE, TAME, methyl acetate). The integration of reaction and separation in one unit operation leads to a reduction of equipment and operational costs as well as to an increase in yield of equilibrium-limited reactions. Enzyme-catalyzed reactions often show a thermodynamic and/or kinetic limitation which can be overcome by this kind of integrated reaction/separation setups. In biocatalytic applications there are different examples for integrated processes, e.g. *in situ* adsorption, *in situ* extraction and reactive membrane processes [Bechtold et al., 2009[1]]. Due to the thermal sensitivity of the enzymes, up to now the reactive distillation was not considered for enzymatic reactions, although improved immobilization methods and the discovery of more robust enzymes would in principle allow the application in reactive distillation processes.

Presented work

In the current work, a silica-based coating was developed for commercial structured packings Montz A3-500 in order to immobilize lipase CALB in reactive distillation columns. The advantage of the coating procedure is that the increased surface area of the packing can be used both for a better vapor-liquid mass transfer and as contact area between substrates and biocatalyst. Furthermore, the coating is produced in the sol-gel process which in principle allows the coating of all kind of commercial column internals. By adjusting the sol-gel parameters the chemical and structural properties of the silica matrix can be altered according to the material used as carrier. In addition, the biocatalytic coating can be adjusted to the favored conditions of the enzyme. Lipase CALB showed for example an increased activity in more hydrophobic coatings made of alkoxysilanes [previously shown by Reetz et al. 1997[2]]. The development of the biocatalytic coating is described as well as the influence of certain constituents on the activity and the stability coating.

Subsequently, the new biocatalytic coating was applied in a transesterification reaction of ethyl butyrate with 1-butanol. The withdrawal of one of the products through the distillate stream led to a clear shift in equilibrium to 98% conversion.

This successful introduction of enzymes in a reactive distillation column opened up a new application field for biocatalytic reactions. In the first place stereo- and enantioselective reactions were considered, that cannot be realized with a chemical catalyst or only in several reaction steps. As an example the kinetic resolution of racemic 2-pentanol was performed in the reactive distillation column. For this reaction it was also demonstrated, that withdrawal of one of the products can shift the equilibrium towards almost complete conversion and the residual mixture of (R)-pentanol and (S)-pentyl ester can be separated in the same column. The conversion and the selectivity in the reactive distillation setup is compared with the reaction in a stirred vessel.

For both reaction systems, the stability of the silica-based coating was investigated by monitoring the weight loss of the coating after each reactive distillation experiment. In addition, the thermal and chemical deactivation of CALB was evaluated for the new reaction system at elevated temperatures up to 60 °C.

In summary, the first realization of an enzyme-catalyzed kinetic resolution reaction in a reactive distillation column shows great potential for the enantioselective synthesis. Furthermore, a new immobilization method is presented for the introduction of enzymes in a reactive distillation column, which shows great flexibility for the application of different enzymes.

Outlook

The process of the kinetic resolution of 2-PeOH in the reactive distillation will be optimized in a simulation-based approach. Therefore, the VLE data for the quaternary mixture 2-pentanol-ethyl butyrate-2-pentyl butyrate-ethanol has to be determined since only limited data was available. The enzymatic reaction can be described by a simple Michaelis-Menten two-substrate kinetic with substrate-surplus inhibition.

The project „Enzyme-catalyzed reactive distillation"(SM 82/9-1) is supported by the German Research Association (DFG).

[1] Bechtold, M. and Panke, S. (2009); In situ Product Recovery Integrated with Biotransformations; CHIMIA International Journal for Chemistry, Volume 63, Number 6, June 2009, pp. 345-348(4)

[2] Reetz, M. et al. (1996); Characterization of hydrophobic sol-gel materials containing entrapped lipases; Journal of Sol-Gel Science and Technology August 1996, Volume 7, Issue 1-2, pp 35-43

(341f) Scaling Down Oscillatory Baffled Reactors

Harvey, A. P., Newcastle University
Phan, A. N., Newcastle University
Eze, V. C., Newcastle University
Abernethy, R. A., Newcastle University
mohd-Rasdi, F.,
Originally presented on: 11/5/2013 17:00:00 - 17:21:00

"Scaling *down* Oscillatory Baffled Reactors"

Adam P. Harvey (adam.harvey@ncl.ac.uk)*, Anh N. Phan (anh.phan@ncl.ac.uk), Valentine Eze (valentine.eze@ncl.ac.uk), Fatimah Mohd Rasdi (f.r.b.mohd-rasdi@ncl.ac.uk), Richard Abernethy (richard.abernethy@ncl.ac.uk)

*Corresponding Author:

Professor Adam P Harvey, School of Chemical Engineering and Advanced Materials (CEAM), Newcastle University, UK NE1 7RU

OVERVIEW

Historically one of the most significant issues for intensified technologies was scale-up. For oscillatory baffled reactors (OBRs) there have been many studies incorporating various aspects of scale-up over the last 20 years. Over the last 8 years, however, one strong feature of OBR research has been instead to scale them *down* to so-called "mesoscale" i.e. to diameters of approximately 5mm. This allows various advantages of oscillatory baffled reactors to be exploited in new ways. For instance, the OBR's ability to accommodate reactions of residence times of the order of hours in a relatively compact reactor with plug flow allows long reactions to be screened or optimised in continuous operation, whilst consuming very low levels of feed and producing little waste. This can be very important for expensive feedstocks or very hazardous waste. Another particular advantage of the "mesoOBR" is that the reactor can at the same time uniformly suspend solid particles. This is difficult to do in other forms of continuous screening reactor.

This paper will give an overview of the development of mesoscale OBRs, explaining their current and potential uses.

DESIGN

The development of mesoOBRs has involved investigation of a number of different baffle designs, including conventional orifice plate designs, central (axial) baffles of various geometries and helical baffles. There are strengths and weaknesses of each design. Helical baffles, for instance, exhibit an "operating window" for plug flow that is approximately an order of magnitude wider than the other designs, allowing an enormous flexibility in operating conditions. The latest compact designs of mesoOBRs are even more compact as they involve serpentine channels. They are designed to maximise residence time per unit area of laboratory bench, to allow "longer" reactions (of the order of a few hours) to be screened.

PROCESS SCREENING

The portfolio of systems that can be screened using this device is gradually being increased in terms of the number of phases present. So far examples of liquid systems, liquid-liquid, liquid-solid and liquid-liquid-solid systems have been assessed. In each case multi steady state screening has been demonstrated, followed by dynamic screening (continuous variation of one input parameter, typically residence time or molar ratio), then multivariable versions of both.

One of the themes of the paper will be the ease of experimentation allowed by such small-scale continuous screening devices. This reduced barrier to experimentation has led to more "adventurous" experimentation. This is illustrated using ongoing in-house case studies on:

(i) The kinetics of biodiesel production: biodiesel is a very well known reaction, but nonetheless the system has been used to identify a previously unrealised set of conditions that allow biodiesel to be produced in a residence time of 2 minutes. This should be contrasted with the conventional commercial residence time of 1-2h. This could be the basis of a 30 – 60 fold reduction in reactor volume.

(ii) The crystallization of L-glutamic acid: a new geometry (tetrahedral) of L-glutamic acid crystal has been identified. The size (sub 20mm) and shape of this new form are more uniform than that achieved by other production methods.

(iii) The kinetics of imine synthesis: the rate constants have been rapidly determined, from dynamic multidimensional screening, using online FTIR.

(341g) Control Strategies for the Spinning Disk Reactor

Ghiasy, D.,
Tham, M., Newcastle University
Boodhoo, K., Newcastle University
Originally presented on: 11/5/2013 17:21:00 - 17:42:00

Spinning Disk Reactors (SDRs) are particularly suited to carrying out reactions with inherently fast kinetics. Although abundant experimental work is available in the literature on intensification aspects of SDRs, in terms of enhancement of reaction rates and improvements in yield and product selectivity, very little has been done on devising appropriate control strategies for the unit. Effective control of SDRs may be challenging due to their fast dynamics and short residence times, for example because the conventional sensors and actuators may be too slow. However, some key features of SDRs may prove to be advantageous from a control point of view. For example, the rotational speed of the disc offers an extra degree of freedom in control system design, since the residence time and mixing intensity, thus conversion, may be controlled by adjusting the rotational velocity, instead of more commonly used methods of varying reactant flow rates. Further, in conventional reactors, mixing limitations may introduce large lags in implementation of the controller's command. On the other hand, the enhanced mixing achieved in a SDR may ensure that the controller's corrective action brings the controlled variable back to the set-point value with practically no delays.

In the present work two test processes, namely neutralization of HCl and NaOH, and precipitation of barium sulphate are chosen to investigate the control aspects of SDRs experimentally. The most commonly used controllers based on PI/PID algorithms implemented in LabVIEW, coupled with commercially available instrumentation, are employed to achieve the control objectives. Set-point tracking performance data show that the pH control of the neutralization process can be successfully achieved using a PID controller which manipulates the flow rate of the base stream to the SDR. Addition of a disturbance observer scheme results in further enhancement of the control performance by suppressing the undesired effects of pH system nonlinearity. The conductivity control of the precipitation process is also successfully achieved by manipulating the disc rotational speed, which presents a vastly appealing potential for adopting an innovative approach to process control for SDRs.

Overall, our study indicates that, in order to achieve a given control task in SDRs, it is essential to first develop a comprehensive understanding of the relationship between the manipulated variable and the controlled parameter. It is also required to account for the interaction of different operating parameters which define the hydrodynamics of the thin films. Depending on the complexity of the system it may be unavoidable to resort to more advanced control schemes to achieve a satisfactory performance.

(343a) Improving the Robustness of a Pharmaceutical Crystallization Process

Marchut, A. J., Abbott Laboratories
Nordstrom, F. L., AbbVie
Patel, K.,
Bhagavatula, L., AbbVie
Originally presented on: 11/5/2013 15:15:00 - 15:40:00

Improving the Robustness of a Pharmaceutical Crystallization Process

Alexander Marchut, Fred Nordstrom, Ketan Patel, and Lakshmi Bhagavatula

Process Research and Development (Process Engineering, Solid State Chemistry, and Process Chemistry), Manufacturing Science and Technology

Abbvie

1401 Sheridan Road

North Chicago, IL 60064

Crystallization is often a key component of any pharmaceutical process – the desired active ingredient or intermediate is separated from the solution phase into a solid phase with the desire that impurities remain in the solution phase and the resulting solid particles have reproducible physical properties such as solid form, crystal habit, and particle size. Pharmaceutical crystallization processes must be robust so that the resulting solid phase properties are reproducible and independent of scale. We employed Quality by Design techniques to improve a pharmaceutical crystallization resulting in a more reproducible process with a broad design space. Our changes to the process were subject to certain regulatory constraints which will be discussed.

(343b) Development of An API Micronization Design Space Using a Design of Experiments Approach

Chekal, B. P., Pfizer Global Research & Development
Steeno, G. S., Pfizer Global Research & Development
Originally presented on: 11/5/2013 15:40:00 - 16:05:00

The design space for a micronization process used to reduce the particle size of a low solubility active pharmaceutical ingredient (API) has been developed using a resource-efficient D-optimal design of experiments (DoE) study. Micronization trials were completed using an Isopak-2 SuperJet Microniser (APTM, Switzerland). An experimental plan was designed to examine the impact of three micronization process parameters; feed rate, mill pressure, and Venturi pressure, on the D(v,0.5) and D(v,0.9) particle size averages. A broad range of the parameter space was examined in the DoE study, but with an additional imposed constraint limiting the difference between the mill pressure and the Venturi pressure to guarantee proper micronizer performance. Due to the irregular polygonal shaped parameter space caused by the imposed constraint, a 10-run D-optimal design was constructed to ensure both adequate coverage across the experimental region and precise modeling capability.

Statistical models generated from the measured D(v,0.5) and D(v,0.9) values demonstrated that these particle size averages have a strong dependence on the mill pressure and a slight dependence on the Venturi pressure. The feed rate had no measurable effect on the particle size averages, implying latitude with that parameter. Three subsequent micronization trials were run to independently verify the model predictions for the D(v,0.5) and D(v,0.9) particle size averages. Target ranges for the particle size averages were established with consideration of bioavailability, content uniformity, and drug product manufacturability. Due to its low aqueous solubility, the particle size of the API is limited to D(v,0.5) of no more than 8 microns and D(v,0.9) of no more than 17 microns to ensure bioavailability. By comparing the model predictions with these target ranges, the design space for the micronization process was identified with the constraint of no less than 2.0 bar for the mill pressure.

Use of a D-Optimal DoE for the efficient development of a robust API micronization design space with parameter constraints has been demonstrated as an alternative to traditional micronization development studies.

(343c) Model Based Design of Experiments for Pharmaceutical Process Development

Albrecht, J., Bristol-Myers Squibb Co.
Li, J., Bristol-Myers Squibb Company

Originally presented on: 11/5/2013 16:05:00 - 16:30:00

Design of experiments (DoE's) are widely used experimental methods to build process understanding. However, the classical design of experiments methodology is optimal only for systems with linear responses and where no prior knowledge exists for the system. Classical factorial designs are common within pharmaceutical drug substance process development despite the fact that these processes are non-linear in nature (i.e. temperature, chemical kinetics). Model based design of experiments allows the researcher to maximize, based on any prior knowledge, the information content of a future series of experiments. Before drug substance DoE's are planned, knowledge of the process is available from, at a minimum, route scouting experiments. Even for reactions where the mechanism is not known, credible options can often be proposed from first principles. From the proposed mechanism(s), a family of models can be drafted. Thus, the combination of any amount of data with the experience and judgment of the researcher can be codified into a model more appropriate than the "no prior knowledge" DoE assumption.

A number of model-based experimental design methodologies have been developed that address the needs of experimental design within the pharmaceutical industry. This paper will report model based design of experiments applied to drug substance synthesis and include examples of optimal experimental design ranging from identifying under-explored experimental conditions to Bayesian updating to refine reliable process operating regions.

(343d) Application of in-Situ Raman Spectroscopy in Monitoring the Dynamic Trends of Polymorphic Transformation and Hydrogenation Reaction

Zhang, H., Sunovion Pharmaceuticals Inc.
Prytko, R. J., Sunovion Pharmaceuticals Inc.
Sizensky, M.,
Saranteas, K., Sunovion Pharmaceuticals
Originally presented on: 11/5/2013 16:30:00 - 16:55:00

Application of in-situ Raman spectroscopy in monitoring the dynamic trends of polymorphic transformation and hydrogenation reaction

Haitao Zhang, Bob Prytko, Mike Sizensky, Kostas Saranteas

Chemical Process Research & Development, Sunovion Pharmaceuticals Inc., 84 Waterford Drive, Marlborough, Massachusetts 01752, U.S.A.

Abstract

During the last decade many novel manufacturing practices such as Process Analytical Technology (PAT), Quality by Design (QbD), and Real Time Release (RTR) have been introduced to the pharmaceutical industry, and they are expected to be an effective aid to pharmaceutical manufacturing. On the basis of a recent FDA directive, this mode of operation will transform the manufacturing aspect of the pharmaceutical industry.

In crystallization processes, control and optimization of the polymorph formation requires in-situ quantitative measurement of the concentration of different polymorphs. Herein, we report the application of in-line Raman spectroscopy to monitor the polymorphic transformation of DSP-6373 during crystallization process. Kinetically preferred Form II of DSP-6373 was transformed into thermodynamically stable Form I during a solvent-mediated phase transition in isopropanol. The feasibility of identifying the differences between the two polymorphs has been demonstrated by collecting and comparing the Raman spectra of the pure polymorphic crystals. A calibration curve has been built based on different ratios of the polymorphs. The in-line Raman spectroscopy can effectively identify the two forms of crystals in solution as well, which allows it to be used to capture the dynamic trend of the polymorphic transformation.

This study experimentally demonstrates the advantages of using in-line Raman spectroscopy for monitoring in-situ pharmaceutical crystallization by detecting the polymorphic transformation, which is fundamental in the development and operation of industrial crystallization processes. A second application of in-line Raman spectroscopy technology was also explored for the hydrogenation reaction of an enacetamide to an acetamide in order to monitor the real time performance of the reaction and obtain in-line kinetic data.

(343e) Application of Combined Experimental and Modeling Approach to Understand pH Driven Reactive Crystallization

Vaidyaraman, S., Eli Lilly and Company
Conder, E., Eli Lilly & Co
Brown, G., Eli Lilly and Company
Stack, A.,
Phillips, M., Eli Lilly & Co
Viswanath, S., Eli Lilly and Company
Moher, E., Eli Lilly & Co.
Originally presented on: 11/5/2013 16:55:00 - 17:20:00

Application of Combined Experimental and Modeling Approach to Understand pH driven Reactive Crystallization

Authors: Shankar Vaidyaraman, Edward Conder, Gregory Brown, Angela Stack, Michael Phillips, Shekhar Viswanath, Eric Moher

Small Molecule Design and Development, Eli Lilly and Company, Indianapolis, IN 46285

This talk will discuss a combined experimental and modeling approach to understand pH driven reactive crystallization. A mechanistic approach was taken by using experiments to determine the fundamental parameters of the system (pKa, solubility product, pseudo-growth constant). A "fit for use" mechanistic model accounting for the different reaction equilibria in solution was used to link the fundamental parameters to process parameters (addition type, pH during crystallization, time duration of crystallization etc.). The model was used to help plan experiments to understand the impact of flow upsets on pH excursion during crystallization. The model was also used to generate a response surface of pH vs other process parameters to provide guidance to pH controller design.

(343f) Model Based Development and Validation of a Preparative Size Exclusion Chromatography Step

Hansen, E. B., Novo Nordisk
Hansen, T. B., Novo Nordisk A/S
Sejergaard, L., Novo Nordisk
Originally presented on: 11/5/2013 17:20:00 - 17:45:00

Model based development and validation of a preparative Size Exclusion Chromatography step

Ernst B. Hansen, Thomas B. Hansen and Lars Sejergaard Novo Nordisk A/S, Hagedornsvej 1, DK-2820 Gentofte, Denmark

The application of mechanistic models for process development and optimization has been a topic of discussion in the Pharmaceutical industry for many years. Mechanistic models are considered as an alternative to a standard DoE type empiric approach. Most often the DoE approach is preferred based on a cost-benefit evaluation where the time consumption and the complexity are found to extensive when using mechanistic modelling. In our company we use mechanistic models for the development of selected downstream process steps. A Model based approach is preferred when a suitable model has been established, in the literature or in-house, and the process step is considered a key step regarding economy and/or product quality.

The presented case deals with Size Exclusion Chromatography (SEC) as the final purification step for a therapeutic protein. The model was developed with the purpose of using simulation for the process evaluation which is done in preparation to the production scale *Process Performance Qualification* (PPQ). The main function of the SEC step is to reduce the content of larger proteins (HMWP and dimer). The model is based on differential mass balances for each component and includes axial dispersion and mass transfer as dispersion effects. The model was calibrated using four lab runs with variation in load, flow and bed height. The fitted model parameters for each component were the exclusion coefficient ($0<K_d<1$) which determines the retention in the column and the mass transfer coefficient related to dispersion (peak width). Before doing the actual process evaluation we had to deal with a challenge concerning the concentration in the product pool from the SEC step. The feed concentration to the SEC step varied quite a bit and since the SEC step can only dilute the product (no adsorption takes place) the variation in the feed concentration results in a variation in the pool concentration. As this is the final step before formulation there is a lower limit to the concentration which must be met. Using simulation we were able to create enough data to visualise the problem and come up with a new strategy for loading the column which can compensate for the variation in the feed concentration. The *process evaluation* is a systematic check of the performance of the step when varying the important process parameters within their operating range. Due to the experimental constraints (time, availability of product) this is often done as a reduced design (DoE). Since we were using simulation there was no need to set up a reduced design. The evaluation was thus based on 216 simulations covering the operating range of the important process parameters (e.g. load, flow, bed height, OD cut). A perhaps overlooked advantage of using simulation is that both the concentration and the content (impurity levels) of the feed solution can be varied as part of the study without any extra effort such as dilution, concentration, or spiking with impurities.

Our experience of using modelling is very positive. The initial extra energy put into understanding and developing the models and the methods for calibration is worthwhile. But, contrary to the general discussion of which approach is better, it is not a reduction in resources or time consumption that drives us. It is the level of process understanding that we gain from using a model based approach compared to an empiric approach. Knowing what is going on we raise the bar for what is expected and what is unexpected. When the model describes the data well we are confident that the process is running as would be expected. This can include non-linear and counter intuitive behaviour. On the other hand when there is a systematic descrepancy between the model and the data we acknowledge that something unusual is going on and start looking for the unexpected phenomena.

(385b) Use of Single-Pass Tangential Flow Filtration for Decoupling Upstream and Downstream Processing

Brinkmann, A., Biogen Idec
Westoby, M. M., Biogen Idec
Elouafiq, S., Biogen Idec
Originally presented on: 11/5/2013 15:40:00 - 16:05:00

Higher cell culture productivity has led to a bottleneck during downstream purification. Various ways have been explored to decouple the upstream and downstream processes to allow for more flexible processing and enable the more efficient use of existing facilities. In this presentation, we will explore the feasibility of using tangential flow filtration to provide decoupling by concentrating and freezing centrate. However, traditional fed-batch tangential flow ultrafiltration (TFF) can be limited due to the size and minimum working volume of TFF systems. Single-pass tangential flow ultrafiltration (SPTFF) provides a viable solution by providing high concentration factors without the need of a recirculation vessel. This reduces overall system size and hold-up and enables concentration in-line with other operations. The SPTFF process was developed at bench scale to determine the appropriate operationg conditions for the desired concentration facter, and then demonstrated at pilot scale to show scalability and evaluate the effect on the downstream purification steps and product quality.

(385c) Clarification of Succinic Acid Fermentation Broth By Ultrafiltration and the Membrane Fouling Mechanism

Wang, C., Institute of Process Engineering, Chinese Academy of Sciences
Wan, Y., Institute of Process Engineering, Chinese Academy Of Sciences
Xing, J., Institute of Process Engineering, Chinese Academy of Sciences
Originally presented on: 11/5/2013 16:05:00 - 16:30:00

Ultrafiltration was investigated to clarify succinic acid fermentation broth in consideration of integrating fermentation and separation process and removal the product in situ. Different membranes(PES 100kDa, PES 30kDa, PES 10kDa and RC 10kDa) were used and two models were applied to analyze the fouling mechanism. Results indicate that ultrafiltration is feasible in clarifying succinic acid fermentation broth. Almost all the microorganism cells (99.6%) were removed from the fermentation broth. Proteins were also removed effectively by all membranes selected in this study. The removal rate was 79.86% for PES 100kDa, 86.43% for PES 30kDa, 86.83% for PES 10kDa, and 80.06% for the RC 10kDa. After ultrafiltration, clearer permeate was obtained comparing with the centrifugation. Among all tested membrane, PES 10kDa gave the best results which showed relative higher initial flux (18.86 L/m2 h), highest protein removal rate (87%), and less flux drop rate. Resistance-in-series model was applied to determine the main factor that caused the operation resistance. Results showed that most membranes selected in this study tended to be fouled by cake layer or concentration polarization. Hermia's model, which is composed of four individual sub-models, was used to analyze the predominant fouling mechanism for the used membranes. Results showed that the fouling of RC 10kDa and PES 30kDa was controlled by the complete blocking mechanism, while PES 100kDa was controlled by the intermediate blocking and PES 10kDa was controlled by cake layer. This conclusion was also proved by SEM photos. Membrane characteristics were analyzed before and after ultrafiltration by AFM and goniometer. Both contact angle and roughness of most membranes increased after ultrafiltration.

(385d) Effect of Filtration On the Concentration of Aqueous Ions in Process Water and Filter Cake Moisture During Iron Ore Beneficiation

Haselhuhn, H. III,
Kawatra, S. K., Michigan Technological University
Halt, J., Michigan Tech University
Originally presented on: 11/5/2013 16:30:00 - 16:55:00

The filtration processes within selective flocculation and dispersion iron ore beneficiation plants cause aqueous ions from the process water to concentrate within the filter cake moisture. These high concentrations of ions can lead to significant operational issues during downstream pelletization and reduction processes. This paper examines the ionic content of concentrated iron ore slurry that is filtered at an operating selective flocculation and dispersion hematite beneficiation plant and relates it to the ionic content of the moisture in the resultant filter cake. Water analyses were conducted to determine the sodium, potassium, calcium, magnesium, fluoride, chloride, nitrate, nitrite, carbonate, sulfate, and phosphate concentrations within the filter feed slurry and filter cake moisture of an operating hematite beneficiation plant. Results have shown that all tested ions within the process are more concentrated within the filter cake moisture than in the filter feed slurry. This has been attributed to the physical and chemical surface interactions of these ions with the iron ore particle surfaces. Of these ions, phosphorus concentrates the most with the filter cake moisture content having a phosphorus content that is 24.67 times that of the filter feed slurry.

(412a) An Integrated Strategy for Production of Liquid Transportation Fuels Via Simultaneous Catalytic Conversion of Cellulose and Hemicellulose From Lignocellulosic Biomass

Han, J.,
Sen, S. M., University of Wisconsin-Madison
Alonso, D. M., University of Wisconsin-Madison
Dumesic, J. A., University of Wisconsin-Madison
Maravelias, C. T., University of Wisconsin-Madison
Originally presented on: 11/6/2013 8:30:00 - 8:55:00

To date, several studies have focused on the development of processes for the catalytic conversion of the cellulose fraction of the lignocellulosic biomass. However, there are few studies conducted to develop a process for the simultaneous catalytic conversion of cellulose and hemicellulose fractions to liquid hydrocarbon fuels. Accordingly, our goal in this study is to develop an integrated strategy for the simultaneous catalytic conversion of these fractions. In the proposed strategy, cellulose and hemicellulose are converted to the same product, levulinic acid (LA), using LA-derived γ-valerolactone (GVL) as a solvent.[1] Then, the LA is catalytically reduced to GVL,[2] and subsequently, GVL is converted to a mixture of butene oligomers.[3] To generate the integrated strategy, we develop interconnecting separation systems to achieve experimentally optimized feed concentrations for the catalytic conversion steps.

A potential drawback of this strategy is its high heating requirements, hence we perform heat integration to maximize heat recovery. After heat integration, the remaining heating requirement is satisfied from combustion of biomass residues, which are also used to produce high-pressure steam for electricity generation in a turbogenerator unit (basic design). However, the cost of the turbogenerator unit accounts for a significant fraction (35-40%) of the total capital cost.[4] Thus, we develop an alternative design, in which there is no electricity generation, and electricity is externally supplied. In this case, biomass residues are only used to generate steam for heating, and the excess residues are disposed. The basic and alternative designs are evaluated and compared with a lignocellulosic ethanol production process proposed by the National Renewable Energy Laboratory (NREL)[5]. To this end, we select the same feedstock and processing rate (2,000 dry tons of corn stover per day) and the same economic parameters and assumptions.

Based on the capital and operating costs, we perform a discounted cash flow analysis to determine a minimum selling price (MSP) of butene oligomers that makes the net present value of the project equal to zero. We find that the MSP for the basic design is 33¢ per gallon of gasoline equivalent (GGE) higher than the alternative design (\$5.09 GGE^{-1}) since the electricity credits obtained in the basic design is low (1.6% of the total production cost). Although the ethanol process yields a higher fuel production than the catalytic strategy, the MSP for the alternative design is 4¢ per GGE lower than the ethanol process (\$5.13 GGE^{-1}) since the total annualized cost of the alternative design is significantly lower than that of the ethanol process.

Furthermore, we evaluate different feedstock alternatives based on their compositions and prices, since the feedstock cost is the major contributor of the operating costs. We find that the use of hybrid poplar is the best option (MSP of \$4.29 GGE^{-1}) among all feedstock alternatives due to its low price (\$51 dry ton^{-1}) and high sugar content (~65 wt%), which in turn, requires less amount of feedstock to produce same amount of fuel. We obtain even better economics (\$4.01 GGE^{-1}) in case of selling biomass residues instead of disposing or burning residues for electricity generation, assuming that biomass residues can be sold at a price (\$88 ton^{-1}) equivalent to that of low-purity lignin that is used as a fuel. Finally, we carry out sensitivity analyses on various key economic parameters. We observe that the total project investment and the internal rate of return (IRR) have the largest impact on the MSP; a 20% increase in the total project investment and the IRR increases the MSP by 10% and 9.5%, respectively.

References:

1. D. M. Alonso, S. G. Wettstein, M. A. Mellmer, E. I. Gürbüz and J. A. Dumesic, *Energy & Environmental Science*, 2013, **6**, 76-80.
2. S. G. Wettstein, D. M. Alonso, Y. Chong and J. A. Dumesic, *Energy & Environmental Science*, 2012, **5**, 8199-8203.
3. J. Q. Bond, D. M. Alonso, D. Wang, R. M. West and J. A. Dumesic, *Science*, 2010, **327**, 1110-1114.
4. S. M. Sen, C. A. Henao, D. J. Braden, J. A. Dumesic and C. T. Maravelias, *Chem. Eng. Sci.*, 2012, **67**, 57–67.
5. F. K. Kazi, J. A. Fortman, R. P. Anex, D. D. Hsu, A. Aden, A. Dutta and G. G. Kothandaraman, *Fuel*, 2010, **89**, S20–S28.

(412b) " Development of Process and Optimization of Lignocellulosic Biomass for Its Conversion to Sugars for Application in Biofuels , Jet Fuels and Specialty Chemicals"

Parekh, S., Sweet Water Energy

Originally presented on: 11/6/2013 8:55:00 - 9:20:00

Ethanol as a liquid transportation fuel is currently produced from various renewable resources including corn ,wheat and energy crops. However, ethanol production from cellulosic biomass an attractive alternative solution and is environmentally sustainable. Biomass such as plant fibers can be hydrolyzed to yield a variety of sugars that can be subsequently fermented by microbes to yield ethanol. First generation of cellulosic ethanol technologies involved four steps like pretreatment , enzymatic saccharification to derive fermentable sugars, fermentation operation followed by product recovery.

Sweetwater Energy, Rochester ,NY has pioneered a unique biomass pretreatment and bio technology using proprietary sophisticated engineered pretreatment and further optimized which is capable of hydrolyzing biomass to drive sugar readily amendable saccharification of sugars in hemicellulose as well as cellulose fractions to monomeric C5and C6 sugars. The improved SWE biomass conversion process can be used on simple substrates wide variety of lignocelluloses biomass such as corn stover, switch grass, corn cobs and other woody biomass with ability to separate C5 sugar or use as blended(C5and C6) streams while achieving yields of sugar recovery exceeding 86% of theoretical value. This presentation will review SWE technology and overall approach to process development, and will highlight recent advances made at SWE in establishing demo model capable of processing 3 MT/day of cellulosic Biomass and the launch of a commercial process processing 21 million lbs of sugars /year for bolt on operation with the corn plant. The importance of pretreatment conditions, enzyme dosing, and various process strategies enabling C5 and C6 sugar and its refinement and polishing that can then be used for commercial production of ethanol fuel, jet fuels , biochemical and specialty chemicals will be discussed.

(412c) Bio-Oil Production and Catalytic Upgrading to Transportation Fuels

Nan, W., Stony Brook University
Mahajan, D., Brookhaven National Laboratory
Kim, T., Argonne National Laboratory
Krishna, C. R., Brookhaven National Laboratory
Originally presented on: 11/6/2013 9:20:00 - 9:45:00

Abstract

Increasing fossil fuel prices and the demands of clean energy have accelerated research on renewable new energy sources. Bio-oil (also known as pyrolysis oil) which can be derived from lignocellulosic biomass by fast pyrolysis process has the potential to be substitute for petroleum-drived transportation fuels. In addition, with numerous combustion tests, it has been shown that bio-oil can be burned efficiently in internal combustion engine without any major modification. However, due to its high acidity (PH: 2-3), high oxygen (30-60wt %) and water (15-30 wt %) contents, bio-oil has lower energy density (15-19 MJ/kg), compared with petroleum (40MJ/kg). Furthermore, bio-oil is thermal unstable that tends to age and results in phase separation at room temperature. Therefore, upgrading of bio-oil is necessary before it can be used as a transportation fuel.

The high oxygen content of bio-oil imparts undesirable properties. To upgrade bio-oil, several metal catalysts such as $NiCl_{2,}$ Ru, Rh and $CoCl_2$ supported on γ-Al_2O_3, C and SiO_2, slurried in polyethylene glycol solvent, were evaluated at 150 - 270°C and in the presence of H_2, bio-oil was deoxygenated by a combination of hydrodeoxygenation and decarboxylation reactions while oxygen is removed as H_2O and/or CO_2. Gas analysis of the reactor yielded: CO_2 (11%) and CH_4 (0.84%), indicating that some oxygen was efficiently removed from the bio-oil partly. The GC-MS analysis showed that propylene glycol and hexadecane were produced as the major products.

(412d) A Techno-Economic Comparison of the Biobased p-Xylene Production

Lin, Z., Rutgers, The State University of New Jersey
Nikolakis, V., University of Delaware
Ierapetritou, M. G., Rutgers, The State University of New Jersey
Originally presented on: 11/6/2013 9:45:00 - 10:10:00

A techno-economic comparison of the biobased p-xylene production

Zhaojia Lin[1], Vladimiros Nikolakis[2], Marianthi Ierapetritou[1]

[1] Department of Chemical and Biochemical Engineering, Rutgers - The State University of New Jersey

[2.] Catalysis Center for Energy Innovation & Department of Chemical & Biomolecular Engineering, University of Delaware

This work focuses on a techno-economic comparison of the biobased p-xylene production. Biomass is the only potential renewable alternative to petroleum of the production of high-value and low-volume chemicals [1]. The development of efficient and economically sustainable routes for biobased p-xylene production has raised a lot of interest since p-xylene is widely used as a raw material to produce polyester polyethylene terephthalate (PET) -- a polymer resin broadly used in the synthesis of fibers, films, and beverage containers [2]. In this work we focus on the techno-economic comparison of two alternative productions of biobased p-xylene based on the process patented by Gevo Inc. [3, 4] and the discoveries from CCEI (Catalysis Center for Energy Innovation) in this work [5-7].

Gevo's invention involves the fermentation of biomass raw materials (starch) to iso-butanol and then isobutanol conversion to p-xylene through dehydration, oligermerization and dehydrocyclization [4, 8]. Another alternative process is based on CCEI discoveries is to form p-xylene via 5-hydroxymethyfurfural (HMF) and 2,5-dimethylfuran (DMF). Particularly, a biobased p-xylene production process can be envisioned to use biomass (starch) as starting raw material that upon depolymerization and isomerization produces fructose to be dehydrated to HMF. HMF is then hydrodeoxygenated to DMF using hydrogen and DMF can react with ethylene to form p-xylene [5-7, 9].

Those two conversion paths have different advantages and disadvantages in terms of production and purification processes. In comparison to the maturity of petrochemical-based chemical production processes after the last few decades' development, biobased chemical productions are still at early stage and the knowledge of reaction and purification process used in the production of biobased products is limited. Thus, techno-economic analysis is a useful tool to assess the feasibilities and to compare the economics of the alternatives of the production pathways. Sensitivity analysis is performed to address the impacts of uncertainties of economic parameters such as the plant capacity, the costs of raw materials and catalyst cost, and the most significant reaction factors of the technological development, i.e. conversion and selectivity of the involved reactions.

The production process is simulated in Aspen Plus® and the economics analysis is conducted in Aspen Economic Analyzer®. The main findings contain the minimum p-xylene selling prices of both GEVO and CCEI processes. The minimum selling prices of both processes are comparable though much higher than petroleum-based p-xylene from naphtha reforming. Biomass raw material starch is a big contribution for both processes. In Gevo's process, fermentation-related cost such as enzyme and biocatalyst is another big fraction; whereas CCEI requires other raw materials such as ethylene and hydrogen which contribute a big fraction. Selectivity is always favored to improve economics.

References:

1. Bozell, J.J., *Feedstocks for the Future – Biorefinery Production of Chemicals from Renewable Carbon.* CLEAN – Soil, Air, Water, 2008. **36**(8): p. 641-647.

2. Chadwick, S.S., *Ullmann's Encyclopedia of Industrial Chemistry.* 1988(A21): p. 235-237.

3. Peters, M.W.T., Joshua D.; Jenni, Madeline; Manzer, Leo E.; Henton, David E. , *Integrated process to selectively convert renewable isobutanol to p-xylene*. 2011, US20110087000A1.

4. Peters, M.W., Taylor, Joshua D., Jenni, Madeline, Manzer, Leo E., Henton, David E., *Integrated process to selectively convert renewable isobutanol to p-xylene*. 2011, Gevo Inc.

5. Williams, C.L., et al., *Cycloaddition of Biomass-Derived Furans for Catalytic Production of Renewable p-Xylene.* ACS Catalysis, 2012: p. 935-939.

6. Nikolla, E., et al., *"One-Pot" Synthesis of 5-(Hydroxymethyl)furfural from Carbohydrates using Tin-Beta Zeolite.* ACS Catalysis, 2011. **1**(4): p. 408-410.

7. Roman-Leshkov, Y., et al., *Production of dimethylfuran for liquid fuels from biomass-derived carbohydrates.* Nature, 2007. **447**(7147): p. 982-985.

8. William A. Evanko, A.M.E., David A Glassner, Aristos A. Aristidou, Kent Evans, Patrick R. Gruber, Andrew C. Hawkins, *Recovery of higher alchools from dilute aqueous solutions*. 2009, Gevo Inc.

9. Lin, Z., M. Ierapetritou, and V. Nikolakis, *Aromatics from Lignocellulosic Biomass: Economic Analysis of the Production of p-Xylene from 5-Hydroxymethylfurfural.* AIChE Journal, 2013: p. n/a-n/a.

(412e) Scale-Up Evaluation of Ionic Liquid Based Sugar Production At High Solid Loading

Li, C., Lawrence Berkeley National Laboratory
Tanjore, D., University of California-Riverside
Wong, J., Lawrence Berkeley National Laboratory
He, W., Lawrence Berkeley National Laboratory
Gardner, J., Lawrence Berkeley National Laboratory
Sale, K., Joint BioEnergy Institute, Emeryville, CA 94608 and Sandia National laboratories, Livermore, CA
Baez, J., Lawrence Berkeley National Laboratory
Simmons, B. A., Joint BioEnergy Institute, Emeryville, CA 94608 and Sandia National laboratories, Livermore, CA
Singh, S., Joint BioEnergy Institute, Emeryville, CA 94608 and Sandia National laboratories, Livermore, CA
Originally presented on: 11/6/2013 10:10:00 - 10:35:00

Ionic liquid (IL) pretreatment is receiving significant attention as a potential process that enables fractionation of lignocellulose and produces high yields of fermentable sugars suitable for the production of renewable fuels. However, successful optimization and scale up of IL pretreatment involves challenges, such as high solids loading, biomass handling and transfer, washing of pretreated solids and formation of inhibitors, which are not addressed during the development stages at the small scale in a laboratory environment. As a first in the research community, the Joint BioEnergy Institute (JBEI), in collaboration with the Advanced Biofuels Process Development Unit (ABPDU), a DOE-funded facility that supports academic and industrial entities in scaling their novel biofuels enabling technologies, have performed benchmark studies to resolve key challenges associated with IL pretreatment using 1-ethyl-3-methylimidazolium acetate and subsequent enzymatic saccharification beyond bench scale.

Using switchgrass, eucalyptus and mixture of both as feedstocks, we have successfully executed 600-fold, relative to the bench scale (6L vs 0.01L), scale-up of IL pretreatment and 60-fold (1.5L vs 0.025L) scale up of subsequent enzymatic saccharification at 15% (w/w) biomass loading. The results generated are consistent with those from the small-scale experiments that have been conducted at JBEI and elsewhere, and indicate there are no fundamental issues in terms of performance associated with the scale-up of an IL-based conversion technology. High solids loading during pretreatment and enzymatic hydrolysis offers several industrial advantages, including decreased reactor size, increased sugar titer and hydrolysis rate, and decreased water consumptions, but also disadvantage such as mass transfer limitations caused by high material viscosity, which needs process optimization. Clear enzymatic inhibition was observed when pretreated materials were not washed completely prior to hydrolysis, and thus, the inhibitors generated during pretreatment and washing can persist into hydrolysis and the downstream steps. As a result, optimization and engineering of the enzyme cocktail and design/operation of the pretreatment and post separation system must be realized before a commercially viable process is realized. The knowledge gained from this initial scale-up study is an essential first step in demonstrating the commercial viability of this promising pretreatment technology.

(412f) Techno-Economic Analysis of An Integrated Catalytic Process for Conversion of Lignocellulosic Biomass to Straight Chain Hydrocarbons

Upadhye, A., University of Wisconsin
Olcay, H., Massachusetts Institute of Technology
Huber, G. W., University of Wisconsin
Originally presented on: 11/6/2013 10:35:00 - 11:00:00

Depleting fossil fuel resources, increasing concerns over environmental effects arising from combustion of these fossil fuel resources and growing global energy needs have driven today's research towards renewable energy technologies. Because of its abundance [1-2], biomass is considered as the most appropriate long-term alterative to fossil carbon. However, a large-scale transition to lignocellulosic fuels is constrained by the economic and technological challenges associated with biomass conversion processes to "drop-in" fuels.

Here, we present a technoeconomic analysis of an integrated catalytic approach toward the production of jet fuel components and commodity chemicals from hardwood feedstock. Our technoeconomic analysis is based on experimental results obtained from a three year study funded by DARPA SurfCat. The first step in this process is the conversion of the biomass into an aqueous hemicellulose stream. The aqueous hemicellulose stream is then converted into furfural. The furfural could then be further converted into a petroleum quality feedstock through aldol-condensation, hydrogenation/hydrocycloaddition and hydrodeoxygenation steps. The solid cellulose stream is converted into Levulinic Acid (LA). LA is then hydrogenated to gamma Valero lactone (GVL). GVL is further converted into olefins and then oligomerized to make jet and diesel range hydrocarbons. We summarize the key technical challenges associated with process integration and review the most promising approaches for the production of mixed alkanes from lignocellulose.

We carry out techno-economic analysis of the entire process for a biorefinery processing 1983 MT/day of red maple feedstock. The process would produce 30 million gallons of jet and diesel fuel range hydrocarbons per year along commodity chemicals (such as acetic acid and HMF) as by-products. Laboratory data was used to estimate the material and energy balances for this process. A detailed process flow sheet was obtained using ASPEN and used to size the process equipment. Using a conceptual process design approach developed by Douglas [3], we estimate the overall economics of the process. From this preliminary techno-economic analysis, we calculate minimum selling price of the jet and diesel fuel. Our initial calculations have shown that jet fuel selling price of $4.06 per gallon can be obtained using this process. Through this analysis, we identify cellulose deconstruction to LA and decorboxylation for olefin formation as the most expensive steps in terms of reactor costs. Here, we present detailed mass and energy balances for the entire process and provide directions for future research in this area.

References:

[1] G. W. Huber and B. E. Dale, *Scientific American*, 2009, 301, 52-59.

[2] Perlack, R. D.; Wright, L. L.; Turhollow, A. F.; Graham, R. L.; Stokes, B. J.; Erbach, D. C. *Biomass as Feedstock for a Bioenergy and Bioproducts Industry : The Technical Feasibility of a Billion-Ton Annual Supply*; US Department of Agriculture (USDA) and US Department of Energy (DOE): April 2005.

[3] Douglas JM. Conceptual design of chemical processes: McGraw-Hill; 1988.

(419a) Application of Quantum-Based Descriptors to the Molecular Design of Ionic Liquids

Chen, Q., University of Kansas
Scurto, A. M., University of Kansas
Camarda, K. V., University of Kansas
Originally presented on: 11/6/2013 8:30:00 - 8:52:00

This project incorporates quantum-based descriptors into a computational molecular design framework for the design of novel ionic liquids. Ionic liquids can be tuned for various applications by varying the structure of the cation and anion. However, many feasible cation/anion combinations have never been synthesized, so experimental data is sparse. Therefore, accurate models for property prediction are a fundamental requirement for computational molecular design. While group contribution methods are effective for some properties, other properties require quantum-level information about the molecule. In this project, the conductor-like screening model (COSMO-SAC) is used to calculate the molecular surface charge distribution (sigma profile) from molecular structure information. The surface charge distribution can then be incorporated into thermodynamic property prediction algorithms. These predictions are then applied to optimization-based design algorithms for novel molecules. These optimization problems are black-box MINLPs, which can be solved to near-optimality by stochastic methods, including using the Tabu search algorithm applied in this work. Examples are shown presenting successful designs for ionic liquids in azeotropic separation processes.

(419b) Use of Water-Oil-Surfactant System Phase Behavior Data/Model for Emulsion-Based Chemical Product Design

Mattei, M., Technical University of Denmark
Kontogeorgis, G., Technical University of Denmark
Gani, R., Technical University of Denmark
Originally presented on: 11/6/2013 8:52:00 - 9:14:00

Emulsions are defined as mixtures of two or more liquids that are normally immiscible, where one liquid, the dispersed phase, is dispersed in the other, the continuous phase. These two phases are generally defined as water and oil, where these terms identify any liquid showing affinity with water or alkanes, respectively. This type of mixtures can be formed through mechanical work (most often provided through mixing), since emulsions are naturally unstable. For this reason, surfactants are usually added to the two main constituent of the emulsion to enhance its stability as a macroscopically homogeneous mixture. Therefore, the correct understanding of the phase behavior of water-oil-surfactant systems is necessary in the field of emulsion science and its adoption in a computer-aided product design framework allows a more flexible choice of the optimal formulation, when an emulsified product is required.

Since several variables (the composition of the three components and temperature) are involved in the description of the ternary phase diagram, there are different ways to graphically represent the phase behavior of these systems. One of the common and the most useful with regards to chemical product design, consists of an X-Y diagram, where the surfactant content (usually in weight percentage) is on the X-axis, while temperature is on the Y-axis. The ratio between water and oil is then kept constant. This diagram is sometimes called "fish-diagram", because of the typical shape of its phase boundaries. Through this diagram, it is possible to recognize three distinct regions: a homogeneous single-phase area, describing the microemulsion domain; a two-phase area, describing the macroemulsion domain (divided into water-in-oil and oil-in-water emulsion domains, depending on the temperature); a three-phase area, describing the coexistence of a microemulsion and a microemulsion. Depending on the specific need, it may be necessary to design formulations that fall in the two-phase area, if an emulsified product is wanted. Or, if a microemulsion is needed, the designed formulation must lie in the single-phase area. The three-phase area, on the other hand, is rarely considered for product purposes, since the stability of the system in these conditions is extremely poor.

To be able to draw the fish-diagram, it is necessary to have available phase boundary data as a function of temperature, measured or predicted with an appropriate model. Unfortunately, this kind of data is not readily available and, therefore, the availability of a model able to describe the phase behavior of the involved systems will allow extrapolation both in terms of temperature and chemical structure of the main constituent of the systems. Consequently, the emulsifying agent (surfactant) can be designed. UNIQUAC, UNIFAC and

CPA has been applied to these systems and preliminary results are shown.

The availability of fish-diagrams with different chemicals involved (both oils and surfactants) and at different water-oil ratios, gives fundamental information for chemical-based product design. In particular, the boundaries for the manufacture of water-in-oil and oil-in-water emulsions (the three-phase area need to be entirely below or above the room temperature, respectively), as well as microemulsions (the dependence from the temperature is not crucial, while it is important that the minimum concentration at which the microemulsion domain is identified). The application to reaction engineering is also highlighted, where the identification of the position of the three-phase domain in necessary to detect the optimum conditions for advanced chemical reactions, occurring in a microemulsified environment.

(419c) QM - Camd: Advances in Computer-Aided Molecular Design of Solvents for Reactions

Siougkrou, E.,
Galindo, A., Imperial College London, Centre for Process Systems Engineering
Adjiman, C., Imperial College London
Originally presented on: 11/6/2013 9:14:00 - 9:36:00

QM - CAMD: Advances in Computer-Aided Molecular Design of Solvents for Reactions

Eirini Siougkrou, Amparo Galindo and Claire S. Adjiman

Department of Chemical Engineering, Centre for Process Systems Engineering, Imperial College London, South Kensington Campus, London SW7 2AZ, UK

ABSTRACT

Many liquid-phase reactions take place in a solvent. The role of the solvent is significant, especially in the control of reaction rates and temperature; a change of solvent can change both the rate and the order of a chemical reaction. Thus, the ability to predict the behaviour of a reaction in a solvent and to choose the most appropriate solvent are crucial challenges. In this work a methodology is developed for the design of optimal solvents for reactions combining Quantum Mechanical (QM) calculations with a Computer Aided Molecular Design (CAMD) formulation. In order to limit the number of QM calculations but also retain accuracy, the kriging[1] approach is used. Kriging is a response surface approach originally developed in geostatistics. It has recently attracted a lot of attention in other scientific areas as well as it is an exact extrapolator with a statistical interpretation which makes it stand out from other methods.

In this approach, starting with calculating the rate constant for a small number of solvents using QM, a kriging response surface of the rate constant as a function of seven solvent properties (dielectric constant, acidity, basicity, surface tension, refractive index, halogenicity and aromaticity) is built. Then the kriging predictor is included in the CAMD formulation as the objective function that needs to be maximised and the optimal solvent based on this predictor is designed. The rate constant in the current optimal solvent is calculated from QM and, if it does not agree with the kriging result, it is included in the set of solvents used to construct the kriging surface. The kriging surface is rebuilt and the procedure iterates until convergence between QM and kriging is achieved. Conventional transition state theory and the SMD[2] solvation model are used for the QM calculations. The proposed methodology is applied successfully to two chemical reactions which exhibit large solvent effects and the results are discussed.

References

[1] D. R. Jones, A taxonomy of global optimization methods based on response surfaces, Journal of Global Optimization, 2001, 21, 345-383

[2] A. V. Marenich, C. J. Cramer, D. G. Truhlar, Universal solvation model based on solute electron density and on a continuum model of the solvent defined by the bulk dielectric constant and atomic surface tensions, Journal of Physical Chemistry B, 2009, 113, 6378-6396

(419d) Automated Selection of Products and Chemistry in Biomass Conversion Through Network Generation and Optimization

Marvin, W. A., University of Minnesota
Rangarajan, S., University of Minnesota
Daoutidis, P., University of Minnesota
Originally presented on: 11/6/2013 9:36:00 - 9:58:00

Biomass can be upgraded to a multitude of valuable products using a variety of chemistries. This leads to two related questions – *What's the best product to make from biomass? And, what's the best route to make it?* These questions have to be addressed simultaneously because while state-of-the-art molecule design methods can be used to efficiently identify the ideal compound for a particular application, it may be possible that it cannot be synthesized from biomass using known chemistries. However, identifying the spectrum of biomass-derived compounds is nontrivial because the network of potential reactions that upgrade biomass can be large (>1000) and complex. We have developed an automated method[1] to: (a) generate the network of synthesis steps that convert biomass to all possible products using our automated network generator, RING[2], and (b) identify optimal compounds for desired applications and their synthesis routes within the network.

In this talk, we will focus on two different classes of compounds – fuels and chemicals – that can potentially be obtained from biomass using heterogeneous catalysis. First, we will consider the problem of identifying the optimal mixture of biomass-derived oxygenates and hydrocarbons that can be blended with gasoline. We use three different objectives as descriptors of competing metrics for determining optimal blends: (a) energy loss during the synthesis as a descriptor of capital cost, (b) external utility requirements as a measure of operating cost, and (c) overall catalyst requirements as an indicator of the 'speed' of synthesis. We find that in all three cases several alternative mixtures of biomass-derived compounds can be identified such that the corresponding gasoline-mixture blend satisfies current ASTM standards on heating value, octane rating, vapor pressure, boiling points, and water solubility. We also find that these alternative solutions are typically composed of varying proportions of higher alcohols such as propanol and/or furans such as dimethylfuran. We further find that multiple routes exist that can form a mixture.

Second, we will consider the problem of synthesizing fatty alcohols from biomass such that nonionic surfactants made from these alcohols have desired physical properties. We will show that in addition to selecting the ideal fatty alcohol and potential synthesis routes, we can automatically obtain process information such as operating windows, thermodynamic bottlenecks, and potential coupling of reactions and/or phases.

[1] W. Alex Marvin, Srinivas Rangarajan, and Prodromos Daoutidis. Automated Generation and Optimal Selection of Biofuel-Gasoline Blends and Their Synthesis Routes, Energy & Fuels, In Press (2013).

[2] Srinivas Rangarajan, Aditya Bhan, and Prodromos Daoutidis. Language-oriented rule-based generation and analysis of complex reaction networks: Description of RING, Computers & Chemical Engineering, 45, 114-123, 2012

(419e) Utilization of Multi-Dimensional Characterization in Virtual High-Throughput Screening

Herring, R. III, Auburn University
Eden, M. R., Auburn University
Originally presented on: 11/6/2013 9:58:00 - 10:20:00

The design and selection of chemicals with targeted properties and activities has greatly benefited from the application of quantitative structure activity relationships (QSAR's). This approach involves the regression of various molecular descriptors, such that a property or activity of interest is accurately modeled as a function of the underlying molecular features. The advent of powerful variable selection techniques has allowed for these models to become highly predictive in nature, which can be taken advantage of in the search for optimal candidates meeting certain predetermined property/activity requirements. It has been shown that each property of interest is best characterized by it's own unique combination of molecular descriptors. These descriptors capture widely varying aspects of a molecule's features and, among many ways, can be differentiated by the dimensionality of information they capture. Simple, low-dimensional descriptors such as molecular weight or number of aromatic rings, while easy to calculate, cannot explain more complicated phenomena such as ligand receptor interactions. On the other hand descriptors containing information such as spatial electrostatic character can explain these interactions, but are much more computationally intensive and vulnerable to statistical noise. The use of these QSAR's in a predictive manner towards identifying promising candidate molecules in a large region of chemical space is known as virtual high-throughput screening (VHTS). Used along-side conventional laboratory based HTS methods, VHTS can significantly reduce the time and costs associated with developing novel chemicals and drugs meeting certain property and activity requirements. The inclusion of descriptors with higher dimensionality, specifically 3D descriptors, has proven very successful but still has its limitations. Most often, these descriptors are calculated from the local energy minima of each molecule in a data set, which is not highly representative of the actual conformations found in solution. In response, techniques have been developed to calculate these spatial descriptors from a conformational ensemble, which can include information such as a molecules flexibility and is commonly known as 4D-QSAR. The utilization of these models in a predictive manner can become quite computationally intensive, especially when searching large regions of chemical space, thus necessitating methods to expedite the search. This contribution details an algorithm useful for VHTS in large regions of chemical space with descriptors of varying dimensionality (0D-4D). The approach utilizes concepts in graph theory and employs genetic algorithm to enable the consideration of a large search space subject to multiple property constraints. The signature molecular descriptor is used to compress the molecular information in a canonical form, which reduces memory storage requirements and facilitates an efficient combinatorial search. A case study will be presented to exemplify the applicability of such an approach.

(419f) Generation of Candidate Reactant Structures Using Signature Descriptors

Dev, V., Auburn University
Chemmangattuvalappil, N. G., University of Nottingham - Malaysia
Eden, M. R., Auburn University
Originally presented on: 11/6/2013 10:20:00 - 10:42:00

Maximization of process efficiency through the selection of optimal chemicals has been one of the highlights of recent process design methods that include product design considerations in their algorithms. These integrated process and product design techniques are exhaustive as they do not commit to a specific chemical component until the final step. These techniques, however, are limited to non-reactive systems. A property based algorithm focusing on the product/molecular design aspects in reactive systems in which all reactants are unknown is introduced in this work. The product of the reaction is constrained by property bounds arising from the process to ensure the optimality of the process. An optimization problem is formulated to select the chemical structures that satisfy the property bounds and have the best dominant property. Molecular design is performed by utilizing the recently developed molecular signature descriptors. The molecular signature of a molecule is the systematic coding of constituent atoms that represents the extended valencies of the atoms to a pre-defined height. The signature of a molecule can be expressed as a linear combination of its atomic signatures [1]. Molecular signatures can assist in tracking the changes in the structure of a molecule due to reactions in the system [2]. This is achieved by generating atomic signatures for the target products through the knowledge of their general structural formula. Different qualitative structure activity/property relationships (QSARs/QSPRs) have been applied in the developed algorithm to estimate properties from the molecular structure. QSARs/QSPRs can be expressed in terms of different topological indices while the topological indices of molecules can be represented in terms of molecular signature descriptors [3]. Thus the target properties have been related to the signatures. In the developed method, the problem is formulated in terms of the number of occurrences of each signature, i.e. the number of times each signature appears in the molecular structure. This allows for the dominant property and the associated property constraints to be expressed in terms of the number of occurrences of the signatures in the product molecule. After identifying the occurrence values, the molecular structure of the reaction product is generated using the algorithm developed by N.G. Chemmangattuvalappil [2]. Next, from the chemical equation, the structure of the reactants is determined by considering the appropriate addition and/or removal of groups. Thus, candidate reactant structures can be generated irrespective of the type and number of reactions as the introduced algorithm focuses on the design of products first and then uses the chemical equations to identify the reactant structures. Safety and environmental constraints can also be considered in the set of property constraints. Graph theory principles have been utilized to track the signatures and to avoid the generation of infeasible molecular structures. This contribution will illustrate the developed methods and highlight their use through a case study.

References

[1] D.P. Visco Jr., R.S. Pophale, M.D. Rintoul and J.L. Faulon, 2002, Journal of Molecular Graphics and Modelling, 20, 6, 429-438

[2] N.G. Chemmangattuvalappil, C.B. Roberts & M.R. Eden, 2012, Computer Aided Chemical Engineering, 31, 1356-1360

[3] J.L. Faulon, D. P. Visco Jr. and R. S. Pophale, 2003, J. Chem. Inf. Compt. Sci., 43, 707-720

(419g) New Perspectives On the Integration of Environmental Impacts Within a Molecular Design Framework

Mehrkesh, A., University of Colorado Denver
Karunanithi, A. T., University of Colorado Denver
Originally presented on: 11/6/2013 10:42:00 - 11:04:00

Computer-aided molecular design (CAMD) is a promising approach that has been widely applied towards the design of optimal molecules for a variety of applications. Typically CAMD methods use structure-property models, such as group contribution models, to predict physico-chemical and thermodynamic properties of interest. Apart from toxicity (if GC or QSPR models are available), explicit consideration of other environmental properties is not currently possible in these methods. However, as sustainable product design is increasingly becoming important one needs to consider holistic environmental impacts at the design stage itself. Life-cycle assessment (LCA) is a technique to assess environmental impacts associated with all the stages of a product's life from cradle-to-grave. Currently life cycle impact assessment methods can only be applied after lead candidate molecules have been designed through methods such as CAMD. This can lead to selection of candidates that are sub-optimal interms of LCA indicators such as global warming potential, human health, and aquaitc ecotoxicity. In this presentation we will explore integration of certain aspects of LCA within the CAMD/CAILD design framework for ionic liquids and molecular compound design. Due to the unique nature of formulation of ionic liquid design problem (CAILD) integration of environmental impacts of upstream processes (production phase) is relatively easy while the features of molecular design problem formulation (CAMD) lends itself to easier inclusion of end-of-life (downstream) impacts. Eventhough integration of complete LCA details within CAMD/CAILD framework is currently not possible we believe that inclusion of partial LCA information would lead to more sustainable design of molecules and products.

(438a) Fischer Tropsch Synthesis in Compressed CO2 Atmosphere With Unreacted Gas Recycle

Mondal, K., Southern Illinois University
Sims, A., Southern Illinois University
Soti, M., Southern Illinois University
Carron, D.,
Originally presented on: 11/6/2013 8:30:00 - 8:50:00

The key challenge in applying the Fischer-Tropsch process to produce transportation fuels is to make the capital and production costs economically feasible relative to the comparative cost of existing petroleum resources. To meet this challenge, it is imperative to enhance the CO conversion while maximizing carbon selectivity towards the desired liquid hydrocarbon ranges (i.e. reduction in CH_4 and CO_2 selectivities) at high throughputs. At the same time, it is equally essential to increase the catalyst robustness and longevity without sacrificing catalyst activity. This paper focuses on process development to achieve the above. The paper describes the influence of operating parameters on Fischer Tropsch synthesis (FTS) from coal derived syngas in supercritical carbon dioxide ($ScCO_2$). In addition, the unreacted gas and solvent recycle was incorporated and the effect of unreacted feed recycle was evaluated. It was expected that with the recycle, the feed rate can be increased. The increase in conversion and liquid selectivity accompanied by the production of narrower carbon number distribution in the product suggest that higher flow rates can and should be used when incorporating exit gas recycle. It was observed that this novel process was capable of enhancing the hydrocarbon selectivity, reducing the parasitic loss of carbon efficiency due to carbon dioxide formation during FT synthesis reactions and increasing the life of the catalyst. The use of the supercritical CO_2 ($ScCO_2$) inhibited both methane and CO_2 selectivities while enhancing the rates of synthesis. In addition, the use of supercritical CO_2 was found to prolong the life of the catalyst presumably by removing the heat of reaction from the catalysts surface and solubilizing the waxes that tend to deposit of the surface. The effect of the process conditions on the life of the catalysts will also be presented. Fe based catalysts are known to have a high proclivity for producing CO_2 during FTS. The data of the product spectrum and selectivity on Co and Fe-Co based catalysts as well as those obtained from commercial sources will also be presented. The measurable decision criteria were the increase in CO conversion at H_2:CO ratio of 1:1 (as commonly found in coal gasification product stream) in supercritical phase as compared to gas phase reaction, decrease in CO_2 and CH_4 selectivity, overall liquid product distribution, and finally an increase in the life of the catalysts. Some of the notable findings include negative CO_2 selectivities while obtaining nearly 98 % CO conversion. Increasing the reactor pressure was found to favor longer chain growth. In addition, it is noted that the ratio between CO_2:syngas in the reaction mixture is an important factor in the liquid product distribution. The results from Fe-Zn-K, Fe-Co-Zn-K and Fe-Co-Ru-Zn catalysts on gamma alumina supports will be presented for recycled gas to feed gas flow rate ratio of 0, 0.5, 1, 2, 3, 4 and 5.

(438b) Compact Heat Exchange Reactors for Process Improvement

Vallee, S. J., Chart Energy & Chemicals
Jia, Z., Chart Energy & Chemicals
Originally presented on: 11/6/2013 8:50:00 - 9:10:00

Compact Heat Exchange Reactors for Process Improvement

Steven Vallee, Zhijun Jia

Chart Energy & Chemicals, Inc., 2502 Losey Blvd, La Crosse, WI 54601

Abstract

Compact Heat Exchange Reactors (CHER) have been used in various applications to improve process productivity, efficiency, and safety. CHER combines a compact heat exchanger with a chemical reactor in a single device. The structure consists of multiple alternating layers for reaction and cooling in intimate proximity, with each layer containing multiple mini-channels in the millimeters scale. This type of structure has excellent heat transfer capability and a large surface area to volume ratio that remains constant upon scale-up.

CHER are especially well-suited for many applications with highly exothermic or endothermic reactions, Process Intensification (PI), or mass transfer enhancement (extraction, absorption). This presentation will include a description of CHER and its features, and give multiple examples of the use of CHER to obtain process improvements including fixed-bed Fischer-Tropsch applications.

(438c) Lean NOx Reduction Over LNT-SCR Dual-Layer Catalyst: Impact of SCR Layer

Zheng, Y., University of Houston
Harold, M. P., University of Houston
Luss, D., University of Houston
Originally presented on: 11/6/2013 9:10:00 - 9:30:00

Lean NO_x Reduction over LNT-SCR Dual-Layer Catalyst: Impact of SCR Layer

Yang Zheng, Michael P. Harold*, Dan Luss*

Department of Chemical & Biomolecular Engineering

University of Houston

Houston, TX 77204

Abstract

Diesel NO_x emission control systems combining Lean NO_x Trap (LNT) catalysts with Selective Catalytic Reduction (SCR) catalysts offer potential performance advantages for lean NO_x reduction. The combined LNT+SCR system development goal is to meet the more stringent upcoming EPA Tier 3 emission regulation and/or California's Low Emission Vehicle (*LEV III*) standards. A promising configuration for the combined LNT+SCR system is a dual-layer system, which generates and utilizes NH_3 during lean-rich switching. NH_3 is produced in the underlying LNT layer during the rich phase is captured in the top SCR layer. The stored NH_3 enables additional NO_x reduction during the ensuing lean phase. The main objective of this study is to research and develop LNT-SCR dual-layer catalysts that will achieve high NO_x conversion (> 80%), especially at low temperatures (< 250 °C), while minimizing the precious group metal (PGM) loading, at application relevant space velocities (> 50K h^{-1}). In order to meet this objective we conduct lean-rich cycling for a family of dual layer catalysts with the aim to elucidate the effect of catalyst composition on performance for a range of conditions.

We studied the impact of ceria loading to LNT catalysts on the low-temperature NH_3 yield and NO_x conversion. The addition of ceria to the LNT catalyst used in dual-layer application enables a higher NH_3 yield and NO_x conversion below 250 °C than does the ceria-free LNT [1]. The SCR top layer increases the washcoat diffusion resistance. Meanwhile, NH_3 trapped in the SCR layer is prone to be oxidized to NO_x in the adjacent LNT layer. The rate of this undesirable reaction increases with temperature. Both adverse effects decrease the high temperature performance of the dual-layer catalyst. In order to overcome the architectural drawbacks of the dual-layer catalyst, we studied which type of SCR zoning can increase the low-temperature NO_x conversion while minimizing the decrease in the high-temperature performance. To determine which SCR catalyst is best suited for dual-layer applications, we compared the NH_3 storage capacity and standard SCR activity of Cu-ZSM-5 (10 membered ring) with that of Cu-SSZ-13 (8 membered ring). The use of Cu-SSZ-13 in the dual-layer catalyst improved the low-temperature performance. Finally, an optimized LNT-SCR dual-layer system capable of maximizing the low-temperature enhancement and minimizing high-temperature loss in the cycle-averaged conversion

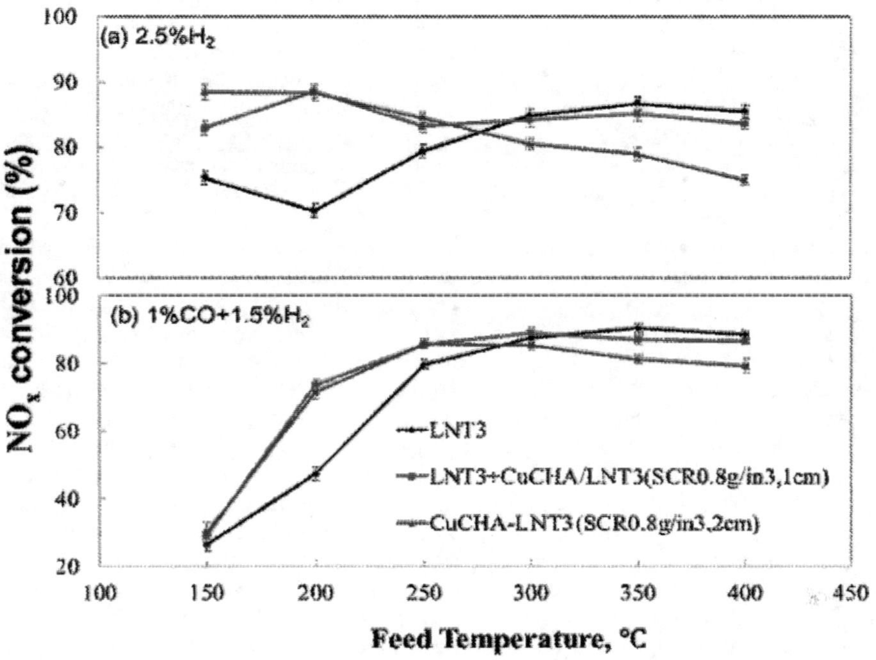

Fig.1 Comparison of NO_x conversion from LNT3, CuCHA-LNT3 (SCR 0.8 g/in^3, 2cm) and LNT3+CuCHA/LNT3 (SCR0.8g/in^3,1cm) using 2.5% CO/H_2 reductant mixtures.

Reference

[1]Y. Liu, Y. Zheng, M.P. Harold, and D. Luss, "Lean NO_x Reduction with H_2 and CO in Dual-layer LNT-SCR Monolithic Catalysts: Impact of Ceria Loading", Top. Catal., **56**, 104Â¨C108 (2013).

(438d) Novel Jet-Loop Reactor for Kinetic Measurements of Gas-Solid Heterogeneous Catalyzed Reactions Involving Commercial-Scale Particles

Nagaraj, A., Texas A&M University-Kingsville

Mills, P. L., Texas A&M University-Kingsville

Originally presented on: 11/6/2013 9:30:00 - 9:50:00

Introduction. Development of new catalytic materials for the selective conversion of next-generation feedstocks to clean fuels and chemicals, or improvement of catalyst compositions for existing processes involving current generation feedstocks, requires reliable kinetic models for design of new reactor configurations or for analysis of transport-kinetic interactions in existing reactor systems (Berty, 1999; Davis and Davis, 2012; Doraiswamy and Uner, 2013; Fogler, 2005; 2010; Froment and Bischoff, 1979; 1990; Roberts, 2008; Salmi *et al.*, 2010; Schmidt, 2004). Procurement of accurate kinetic data for kinetic model discrimination by combining statistical sequential experimental designs with kinetic parameter estimation techniques (Schwaab *et al.*, 2006; 2008) requires a reactor system that can generate the required kinetic equation response variables, such as reaction rates, specie concentrations, and temperatures for a given sequence of user-defined input variables.

Experimental reactors for measurement of reaction kinetics can be broadly classified according to the catalyst forms (powder, granule, or pellet) and the numbers of phases present (gas, gas-liquid, gas-liquid-liquid). A good introductory overview of various reactor designs for both gas-solid catalyzed and gas-liquid-solid catalyzed reactions is provided in standard texts on reaction engineering (*e.g.*, Froment and Bischoff, 1979, Sections 2.3 and 6.5; Levenspiel, 2002, Chapter 21, Section IV), dedicated monographs (Berty, 1999; Chapter 2, Section 2), and review chapters (Pratt, 1987, Chapter 4). It is generally agreed that for the purpose of generating kinetic data, laboratory-scale reactors that operate isothermally and allow direct evaluation of reactant conversions, product yields, and reaction rates from experimental measurements of flow rates and specie concentrations, such as that provided by those where the fluid flow patterns approach perfect back-mixing, are preferred for various practical reasons over systems where the flow patterns approach perfect plug flow (Berty, 1984; 1999). Among various designs with nearly perfect fluid back-mixing, recycle reactors are preferred as important tools for catalyst testing and kinetic studies. They can often be operated at the operating conditions of the commercial-scale reactor, and kinetic rate equations developed from the data they produce can often be used directly for predicting the performance of commercial-scale units. Recycle reactors are also useful for the study of most commercially important reactions, although some exceptions occur, especially for cases where the catalyst activity changes rapidly when compared to typical fluid residence times.

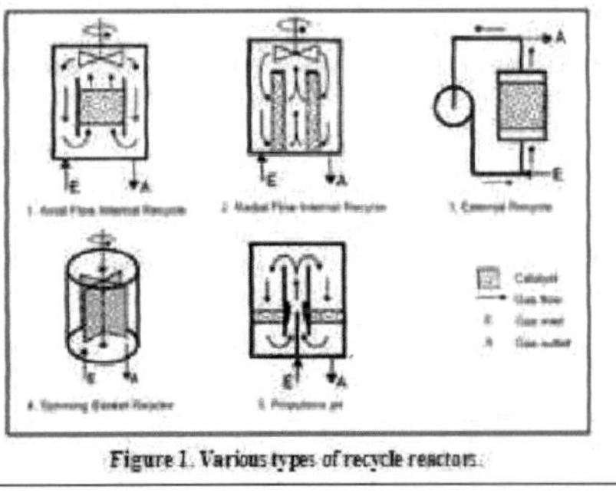

Figure 1. Various types of recycle reactors.

Some examples of various recycle reactor designs that have appeared in the open literature are shown in Figure 1. Types 1, 2, 4, and 5 are *internal recycled* reactors, while Type 3 is an *external recycle* reactor. They are distinguished from each other by the location of the circulating reaction fluid relative to the fixed-bed of catalyst. The various internal recycle reactor designs shown here depend upon a mechanical agitator to create either an axial or radial flow internal circulation of gas. Conversely, the external recycle reactor design uses a compressor to create the required flow rate of circulated gas. One exception here is the so-called propulsive jet internal recycle reactor shown as Type 5, which relies upon a high-speed turbulent jet through an internal draft tube to create the required gas circulation. This latter design does not contain any moving parts, which reduces the complexity and cost when compared to the other designs. This absence of any moving parts is particularly attractive when the reaction kinetics must be evaluated at high temperatures or when using corrosive reaction mixtures since issues associated with bearings and components that would be subject to failure are eliminated.

Objectives. The primary objective of this work is to describe the development, application and modeling of a new experimental reactor for studying the kinetics of gas-solid catalyzed reactions based upon the propulsive jet concept shown above in Figure 1 as Type 5. The propulsive jet concept is more attractive since this approach does not involve any moving parts and is simpler to fabricate. The lack of any moving parts reduces the maintenance requirements, minimizes fabrication costs, and allows for easier set-up and operation. The test reaction used is the oxidation of SO_2 to SO_3 over V_2O_5-based catalysts since increasing emphasis is being placed on development of new catalysts having higher activity at lower temperatures with resulting lower SO_2 emissions for more environmentally-friendly processes. This reaction also represents a challenging test case since previous experience with a conventional Berty-type of reactor resulted in significant issues associated with corrosion of the catalyst basket, excessive bearing wear, and operational interruptions associated with maintenance of the Magne-Drive.

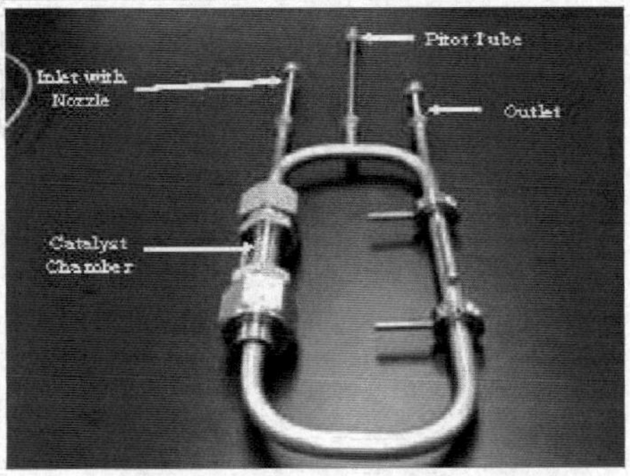

Figure 2. Photograph of new jet loop reactor.

Results and Discussion. A photograph of the new reactor design is shown in Figure 2. The one shown here is constructed of Type 316 stainless steel, although other versions have been constructed from Hastelloy C and other advanced metal alloys. The catalyst chamber is sealed on both ends by SwagelokÔ VCR fittings with metal gaskets for ease of installation. It can accommodate several commercial size catalysts whose largest dimension is ca. 16 mm. The reaction gases are mixed upstream and introduced at a steady flow rate from a gas manifold and removed through the outlet whose back-pressure is maintained constant by a back-pressure regulator. Analysis of SO_2, O_2 and N_2 in the feed and product gases is performed with an on-line GC. A pitot tube located on the return leg is used to measure the internal gas recirculation rate. A large number of experiments were conducted in which the total gas flow rate, gas temperature, reactor total pressure, catalyst particle size, nozzle size, and nozzle location were systematically varied to assess the effect of these variables on the gas recycle rate. The key conclusions from these studies are: (1) nozzle velocity must be maintained above a certain threshold value to generate adequate gas recycle, which was also confirmed by the fluid mechanical calculations; (2) a nozzle I.D. of 0.015-in generates recycle rates between 20 to 45 when gas flow rates between 50 to 950 sccm are used and the nozzle exit is located at an optimal location.

An example of typical kinetic data is provided in Figure 3. These data are based upon temperatures between

400 to 475°C using a gas feed consisting of ca. 1% SO_2, 7.5% O_2 and 82.5% (bal) N_2 at flow rates between 100 sccm to 900 sccm. The resulting rate data were fitted to a rate model that accounts for reversible reaction behavior. The agreement between the reaction rates versus %SO_2 conversion data is shown below in Figure 3a. A parity plot that shows a comparison between the experimental and model-predicted %SO_2 conversions at a selected temperature (425°C) is shown in Figure 3b. Good agreement is obtained which illustrates the validity of a simple but effective reaction rate model. These and other detailed results will be presented that demonstrate the utility of the proposed jet loop reactor as a relatively simple but effective reactor type for studying the kinetics of gas-solid catalyzed reactions.

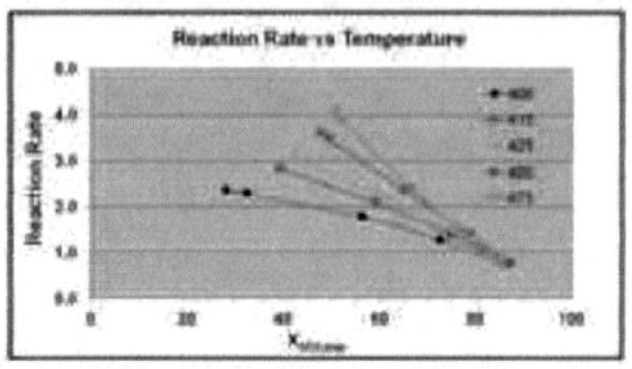

Figure 3a. Reaction rate vs. temperature

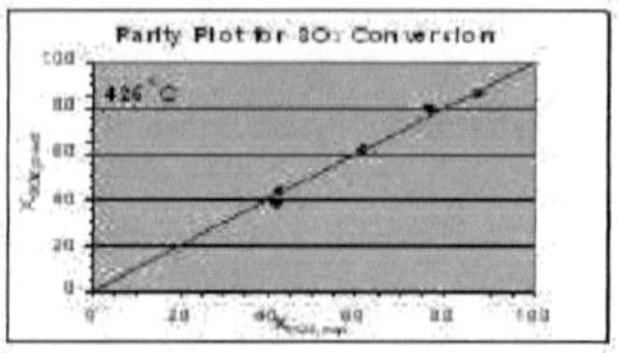

Figure 3b. %SO_2 conversion–exp vs. model

References

Berty, J.M. (1984). 20 Years of Recycle Reactors in Reaction Engineering. *Plant/Operations Progress, 3(3),* 163-168.

Berty, J.M. (Eds.). (1999). *Experiments in Catalytic Reaction Engineering.* Amsterdam: Elsevier.

Davis, M.E., & Davis, R.J. (2012). *Fundamentals of Chemical Reaction Engineering.* New York: Dover Publications.

Doraiswamy, L.K., & Uner, D. (2013). *Chemical Reaction Engineering: Beyond the Fundamentals.* Boca Raton: CRC Press.

Fogler, H.S. (2005). *Elements of Reaction Engineering - 4th Edition.* New Jersey: Prentice Hall.

Fogler, H.S. (2010). *Essentials of Chemical Reaction Engineering.* New Jersey: Prentice Hall.

Froment, G.F., & Bischoff, K.B. (1979). *Chemical Reactor Analysis and Design ? 1st Edition.* New York: John Wiley & Sons.

Froment, G.F., & Bischoff, K.B. (1990). *Chemical Reactor Analysis and Design ? 2nd Edition.* New York:

John Wiley & Sons.

Levenspiel, O. (2002). *The Chemical Reactor Omnibook.* Corvallis: Oregon St Univ Bookstores.

Pratt, K.C. (1987). Small Scale Laboratory Reactors. *Catalysis, 8*, 173-226.

Roberts, G.W. (2008). *Chemical Reactions and Chemical Reactors.* New York: John Wiley & Sons.

Salmi, T.O., Mikkola, J.P., & Warna, J.P. (2010). *Chemical Reaction Engineering and Reactor Technology.* Boca Raton: CRC Press.

Schmidt, L.D. (2004). *The Engineering of Chemical Reactions.* New York: Oxford University Press.

Schwaab, M., Monteiro, J.L., & Pinto, J.C. (2008). Sequential Experimental Design for Model Discrimination Taking into Account the Posterior Covariance Matrix of Differences between Model Predictions. *Chem Eng Sci, 63,* 2408-2419.

Schwaab, M., Silva, F.M., Queipo, C.A., Barreto, A.G., Nele, M., & Pinto, J.C. (2006). A New Approach for Sequential Experimental Design for Model Discrimination. *Chem Eng Sci, 61,* 5791-5806.

(438e) Towards a Better Understanding of Mixing Characteristics of Thin Film Flow in Spinning Disk Reactors

Boodhoo, K., Newcastle University
Mohammadi, S., Newcastle University
Al-Hengari, S., Newcastle University
Originally presented on: 11/6/2013 9:50:00 - 10:10:00

The spinning disc reactor (SDR) is an intensified reactor technology which exploits the benefits of intense fluid dynamics in thin films formed by applying high centrifugal fields to a liquid flowing on the rotating disk surface. The thin films are highly sheared and their surface is covered by numerous unstable ripples which have been shown to promote enhanced heat and mass transfer rates [1,2]. Industrially important reactions such as polymerisation and crystallisation have been performed in the SDR, whereby significant advantages related to improved reaction rates and product properties have been demonstrated [3-5]. In such reaction processes, the mixing characteristics of the thin films are critically important in determining the yield and selectivity of the process as well as the properties of the product(s) formed, especially in fast reaction schemes. Both macro- and micromixing have to be considered for a full characterisation of mixing.

In the present work, a quantitative experimental characterisation of the micro- and macromixing behaviour of thin films formed on the surface of rotating disks is reported. Our studies have demonstrated that the micromixing efficiency of the thin films, quantified via the well-established iodide-iodide test reaction [6], is enhanced under conditions of high disk rotational speeds, high feed flowrates and on grooved disk surfaces. These parameters provide the conditions for the highest shear to prevail in relatively thin continuous films, with a large number of surface ripples for enhanced molecular diffusion. Using a grooved disk with low flowrates can reduce the micromixing efficiency due to rivulet flow resulting from insufficient wetting of the larger surface area. Grooved surfaces are therefore better suited to high throughputs. We have also observed that the feed injection system can have a dramatic influence on the micromixing in the SDR. Multipoint distributors are more effective than single point distributors at high feed flowrates, giving enhanced liquid distribution as one stream feeds into another.

Residence time distribution (RTD) data gathered from online conductivity measurements of the thin films injected with a pulse of potassium chloride as tracer indicate that the higher the disk speed and the liquid flow rates and the lower the viscosity of the processing liquid, the closer conditions within the liquid film on the disk approximate to plug flow behaviour. These effects are attributed to greater degree of turbulence induced in the liquid film, whereby transverse mixing is promoted across the film thickness, resulting in a more uniform velocity profile at any given radial position. The centrifugal force directing the liquid radially to the disc periphery may also play a part in reducing radial dispersion. With flow on a grooved disk, radial dispersion is also observed to be lower than on a smooth disk under identical operating conditions. With the constantly changing topography enabled by the series of concentric grooves across the disk surface, repeated film detachment from and re-attachment to the surface occurs. This is accompanied by induced recirculation or vortices within the film, giving better transverse mixing and eliminating velocity gradients. Based on our experimental data, empirical models relating the Peclet number to the disk operating parameters have been developed for both the smooth and the grooved disks, enabling prediction of plug-flow behaviour for any set of conditions within the range tested.

References

1. A. Aoune, C. Ramshaw, Process intensification: heat and mass transfer characteristics of liquid films on rotating discs, *Int. J. Heat Mass Tran.*, **42**, 2543-2556 (1999).
2. J.R. Burns, R.J.J. Jachuck, Monitoring of CaCO3 production on a spinning disc reactor using conductivity measurements, *AIChE J.*, **51**, 1497-1507 (2005).
3. K.V.K. Boodhoo, W.A.E. Dunk, and R.J. Jachuck, Continuous photo-polymerisation in a novel thin film spinning disc reactor, *in Photoinitiated Polymerisation,* K. Belfield and J. Crivello (Eds.), ACS Symp. Ser. 847, American Chemical Society, Washington D.C., 2003.
4. K.V.K. Boodhoo, R.J. Jachuck, Process intensification: spinning disc reactor for styrene polymerization, *Appl. Therm. Eng.,* **20**, 1127-1146 (2000).
5. L.M. Cafiero, G. Baffi, A. Chianese, R.J. Jachuck, Process intensification: precipitation of barium sulphate using the spinning disc reactor, *Ind. Eng. Chem. Res.*, **41(21)**, 5240- 5246 (2002).
6. Fournier, M.C., L. Falk, and J. Villermaux, A new parallel competing reaction system for assessing micromixing efficiency--Experimental approach. Chem. Eng. Sci., **51(22)**, 5053-5064 (1996).

(438f) Intensified Catalytic Reactor Design for the Conversion of Volatile Fatty Acids Produced By Sewage Treatment Plant

Jayamurthy, M., Brunel University
O'Connell, M., 1Institut fur Mikrotechnik Mainz GmbH
Zapf, R., Institut fur Mikrotechnik Mainz GmbH
Kolb, G., Institut fur Mikrotechnik Mainz GmbH (IMM)
Komodromos, C., Brunel University
Bhattacharya, A., Brunel University
Originally presented on: 11/6/2013 10:10:00 - 10:30:00

Declining petroleum resources and increased demand for fuel are driving force for finding renewable sources of fuel particularly for transport as this sector consumes more than two third of energy produced. Volatile fatty acids (VFA) are produced as intermediates prior to methanogenesis during sewage treatment, and can be sustainably converted to a bio-diesel, since this bio-fuel will not interfere with the food production unlike other biomass based fuels. These VFAs, consisting mostly of acetic, propionic and butyric acids, can be separated and catalytically converted to a diesel-like fuel.

The reaction is exothermic, so heat management is a challenge, whilst efficient mixing of the reactants is critical in obtaining high conversion and selectivity.

An intensified catalytic reactor for the hydrogenation of VFAs has been designed and tested. A new mixing device for a better gas to liquid contact is also included in the design. Ni-Mo, Co-Mo and Pt/Al_2O_3 catalysts have been prepared and coated on metal foams. Both powder and coated catalysts can be tested using this simple reactor design. The Figure shows the reactor components with catalyst coated foams. The reactor is currently being tested at higher pressures upto 30 bar for VFAs conversion; typically reactions have been carried out at 350°C and at ambient pressure. Gas Hourly Space Velocity of hydrogen and VFAs were maintained at 100 h^{-1} and $1h^{-1}$ respectively. The system was tested with a mixture of acetic acid, propionic acid and butyric acid in the ratio 40:40:20. A complete conversion of these mixed acids to hydrocarbons is achieved at the reaction temperature of 350°C. Further studies for optimising conditions are in progress and will be reported in full with a detailed reactor design.

(438g) Actinometric Determination of UV-Radiation in Wet-Tube Electrostatic Precipitator (WESP) and Degradation of Pollutants From Wastewater

Tahir, M. S., University of Gujrat
Saleem, M., Institute of Chemical Engineering and Technology, University of the Punjab, Lahore, Pakistan
Gul, S., University of engineering and Technology
Akhtar, N. A., Natinal Textile University Faisalalabad
Siebenhofer, M., Technical university Graz
Originally presented on: 11/6/2013 10:30:00 - 10:50:00

This paper presents the results of investigation of emission of UV-radiation and the impact of high intensity corona discharge (photon) on actinometric conversions and pollutant degradation with WESP equipped with brush discharge electrodes. Experimental investigations were conducted according to the procedure reported in literature (Rahn et al, 2003; Rahn 1997).Experimental investigations were carried out to determine UV-radiation at various applied voltage (kV) and impact of radiation intensity on chemical conversion of 0.6 M potassium iodide (KI) and 0.1 M potassium iodate (KIO_3) in 0.01 M borax ($Na_2B_4O_7.10H_2O$) for formation of tri-iodide was recorded. WESP equipped with brush discharge electrode produced a suitable intensity of UV-radiation at l = 254 nm. UV-radiation of l = 254 nm was verified using spectrophotometer and validated with the Rahn Actinometer. The capability of WESP for UV-radiation was evaluated. Chemical changes were confirmed with high intensity corona discharge from brush discharge electrodes. Degradation of pollutants (acetone and phenol) in wastewater were confirmed.

(444a) Novel Catalysts and Reactor Concepts for Sustainable Processes: the Use of LCA Data in Process Design

Yaseneva, P.,

Lapkin, A., University of Cambridge

Originally presented on: 11/6/2013 8:30:00 - 8:55:00

Novel catalysts and reactor concepts for sustainable processes: the use of LCA data in process design

Polina Yaseneva and Alexei Lapkin

Department of Chemical Engineering and Biotechnology, University of Cambridge, Cambridge, CB2 3RA, United Kingdom

Abstract

We present a multi-step assessment framework, which was developed to guide decision making process in a process development context. Reaction mass intensity metric, gate-to-gate flowsheet analysis and LCA are used at appropriate stage gates. Specifically results of LCA assessment will be discussed for the case studies of a novel catalyst for water purification process, an intensive process of a pharmaceutical API synthesis and a novel organometallic catalytic process for intermediates synthesis.

Introduction

Assessment of sustainability of novel processes for decision-making purposes is a complex task, as frequently uncertainty of the data precludes the use of full LCA methodology, whereas proxy methods such as mass or energy metrics provide incomplete information. At the same time, information about sustainability of novel technologies is now routinely requested even within research projects. Thus, most EU projects in the area of chemical technologies are accompanied by LCA studies.

Within a large integrated project SYNFLOW (www.synflow.eu) we have developed a three-step approach, using reaction mass efficiency metric for selection of chemical routes, gate-to-gate mass and energy metrics for assessment on flow sheet options and finally LCA for evaluation of the demonstration case studies. These steps coincide with stage-gating within a project: confidence in chemistry to be taken up for process development, evaluation of process options with a focus on clean technologies (novel solvents, process intensification) and finally the detailed analysis of complete process at the stage of demonstration with LCA, targeting detailed flowsheet optimisation.

Here we present results of LCA studies for three specific examples. We show LCA of a novel carbon nanotube-based catalyst for water purification, focusing on comparison of the manufacture of the catalyst, a study of a novel flow-process of stoichiometric reduction of artemisinin, leading to manufacture of an important pharmaceutical API (focusing on comparison of batch vs continuous process) and finally a study of a novel catalytic Buchwald-Hartwig amination under overall flow conditions.

Results and Discussion

Carbon nanotube catalyst for water purification

A highly active catalyst was developed for reduction of an emergent aqueous pollutant, bromate, based on 0.3 %wt Pd deposited onto carbon nanotubes, grown on sintered metal fibre. The TOF of the catalyst was found to be an order of magnitude higher than that of a conventional catalyst based on alumina support (Pd/Al_2O_3). However, there remained the question whether high-temperature CVD process used in the manufacture of carbon nanotubes resulted in a cleaner overall process. The LCA study focused on the manufacture of the two catalysts. In this specific case all environmental impacts for the new catalysts were lower. Detailed analysis of contributions of the individual manufacturing steps to various impacts, e.g. analysis of contributions to cumulative energy demand shown in Figure 1, allows to reveal further options for optimisation of the novel catalyst manufacture [1].

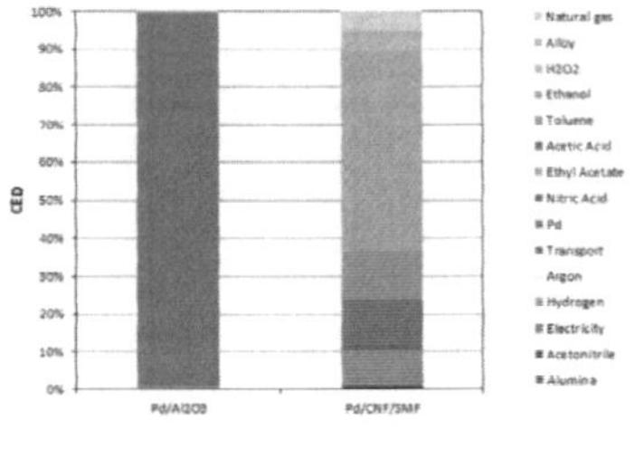

Figure 1. Individual contributions to cumulative energy demand (CED) of the manufacture of the two catalysts.

Artemisinin reduction under intensive flow conditions

We have developed a flow process for reduction of artemisinin to overcome the limitations of a conventional batch process [2]. In the current study we extended the reaction to the final step in the manufacture of the antimalarial API, but also performed an LCA study to compare the novel flow process with the conventional batch reaction [3]. Analysis of LCA results revealed that the solvent used in the manufacture and storage of the reducing agent required for the flow process is having a massive negative impact, see Figure 2. It is becoming critically important to replace the solvent used in a particular stage of the overall process. In this case LCA is instrumental in identifying the critical stage of the overall process that requires optimisation.

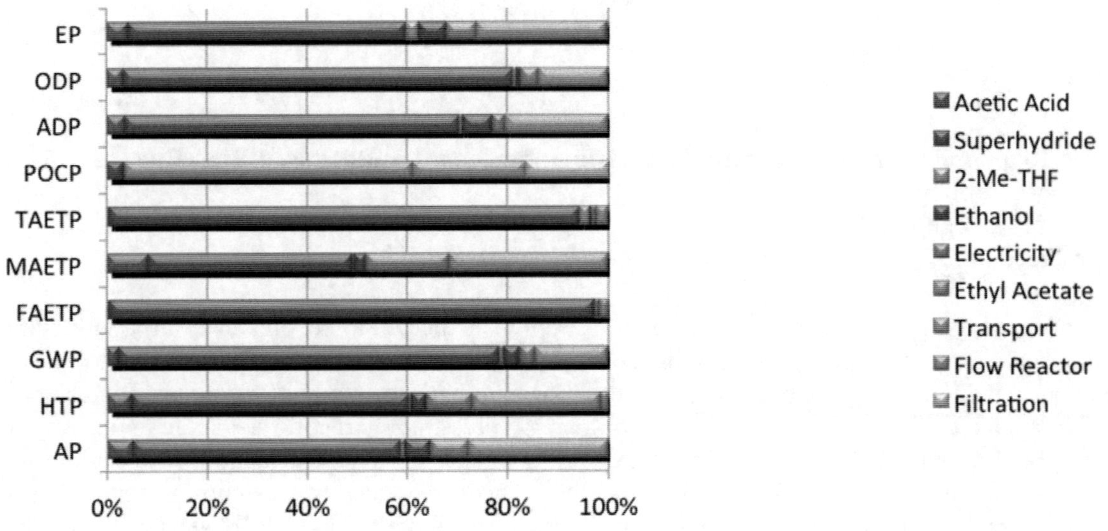

Figure 2. Individual contributions to environmental impacts in artemisinin reduction under flow conditions.

Buchwald-Hartwig amination in flow

The three-step decision making process is illustrated on the example of Buchwald-Hartwig amination. Reaction mass efficiency, process efficiency and LCA are evaluated for this reaction at the different stages in the process design. This approach complements the parallel chemistry-process design methodology being developed within Synflow project, ultimately aiming at a significant reduction of time required for commercialisation of novel technologies, which are also sustainable. Thus, for the initial reaction options we have used RME metric to evaluate the importance of solvent and catalyst recycle. The new process is currently being scaled-up to a demonstrator at an Invite facility of Bayer. We will show the results of stage-gating metrics and initial LCA data.

[1] P. Yaseneva et al., Reduction of bromates on carbon nanofibre supported structured Pd catalysts: experimental and life cycle assessment study, Submitted.

[2] X. Fan, V. Sans, P. Yaseneva, D. Plaza, J.M.J. Williams, A. Lapkin, Facile stoichiometric reductions in flow:

example of artemisinin, Org Proc Res Dev 16 (2012) 1039-1042.

[3] P. Yaseneva et al., Optimisation of the flow process of manufacture of artemether and its LCA study, Submitted.

Acknowledgement: This work was supported by the European Community's Seventh Framework Programme under project "SYNFLOW", NMP2-LA-2010-246461.

(444b) Techno-Economic and Environmental Assessment of Ionic Liquid-Based Biorefineries

Giarola, S., Imperial College London
Shah, N., Imperial College London
Originally presented on: 11/6/2013 8:55:00 - 9:20:00

Although the peak oil is still a debated issue, petroleum supply is expected to become ever more vulnerable of interruptions and current economic system reliance on fossils is exposing all economic sectors to unsustainable price fluctuations [1]. Concerns of maintaining an adequate supply of resources, such as fuels, food, raw materials for a growing population, are also raising a pressing interest for a change in our way to conceive energy and resource provision system [2]. A paradigm shift in the industrial sector from today's large fossils exploitation towards more sustainable practices is also advocated to face current trend of carbon dioxide emissions production, which may soon be approaching greenhouse gas (GHG) levels dangerously close to tipping points of an accelerated global warming [3]. Thus, the future challenge to address the growing world energy demand may be met as long as the supply of higher levels of resources and energy is achieved in a sustainable way promoting low-carbon technologies and renewables [4]. In this respect, the spread of biomass exploitation, as a unique source that can supply both energy and carbon for the production of the commodity chemicals and products, has the potentials to lead to the settlement of a new biomass-based economy [5-6].

Biomass is an abundant, renewable and carbon neutral resource, which is currently significantly exploited for biofuels production from edible components of food crops. However, the major challenge for a future bio-based economy would very likely rely on lignocellulosic feedstock exploitation, being more abundant and less expensive than starchy crops as well as avoiding any direct conflicts with the food chain [7]. Industrial biorefineries have been identified as the most promising route to the creation of a new domestic biobased industry, where biomass conversion processes into fuels, power, and chemicals are integrated through an efficient combination of technologies, to minimise the environmental footprints and ensure the sustainability of all products generated. By producing multiple products, a biorefinery can take advantage of the differences in biomass components and intermediates and maximise the value derived from the biomass feedstock [8]. However, cost-effective technologies for processing lignocellulosic materials into fuels and chemicals are not commercially available. In this context, a central role in determining the overall system profitability is played by the pretreatment unit, where biomass is deconstructed and the three main components (i.e. cellulose, hemicellulose and lignin) are made accessible to subsequent processes [9].

Since the study of Wilkes and Zaworotko [10], ionic liquids have become increasingly popular as reaction and extraction media. The property of ionic liquids which has led to them being classed as green solvents is their very low volatility, thus they do not emit volatile organic compounds and are suitable for reactions at atmospheric pressure. Recently, ionic liquids have begun to be investigated for biomass related processes, yet there are many technological applications and large commercial interests in the field [2]. The use of ionic liquids as lignocellulose deconstruction solvents is also extremely promising [11]. Among the potential applications, ionic liquid can be used to decrystallise cellulose and disrupt the lignin-hemicellulose network. The possibility of removing lignin with ionic liquids and recovering it as precursor for commercial products is also an interesting option [11]. However, deconstruction with ionic liquid will only be viable if some drawbacks are overcome, in particular the cost of ionic liquids. Process design and optimisation is required in order to have high solvent recovery and recycle, but also consumables use reduction, energy integration and waste minimisation. Economic aspects should also be coupled with the concept of ecological sustainability during the process optimisation strategy [12]. Process system engineering techniques play a major role in achieving all the above targets.

This work proposes an investigation of techno-economic performance of a biorefinery based on ionic liquid-pretreatment. The analysis, based on mathematical modelling and process design techniques, aims at identifying the key physical phenomena, the main cost centres affecting the process economics and proposing strategies to develop more efficient biochemical conversion routes for the biomass. This is performed along with the system environmental performance appraisal based on Life Cycle Assessment (LCA) techniques [13]. The final purpose is to provide a mathematical tool for an in-depth economic and environmental analysis of the ionic liquids fractionation, identifying the process enhancements alternatives and outlining potential synthetic routes to biocommodities.

References:

1. Lutz, C., U. Lehr, K. S. Wiebe (2012). Economic effects of peak oil. Energy Policy, 48, 829-834.
2. Tan, S. S. Y., D. R. MacFarlane (2009). Ionic liquids in Biomass Processing. Topics in current chemistry, 290, 311-340.
3. Foley, J. A. (2005). Tipping Points in the Tundra. Science, 310, 627-628.
4. International Energy Agency (2012). World Energy outlook. IEA, Paris (FR).
5. Clark, J. H., F. E. I. Deswarte, editors (2008). Introduction to Chemicals from Biomass Padstow: John Wiley & Sons, Ltd.
6. Ragauskas, A. J., C. K. Williams, B. H. Davison, G. Britovsek, J. Cairney, C. A. Eckert, et al. (2006). The Path Forward for Biofuels and Biomaterials. Science, 311, 484-489.
7. The Royal Society (2008). Sustainable biofuels: prospects and challenges. The Royal Society, London (UK). www.royalsociety.org
8. National Renewable Energy Laboratories (2009). Biomass Program. U.S. Department of Energy, Golden (CO).
9. Aden, A., M. Ruth, K. Ibsen, J. Jechura, K. Neeves, J. Sheehan, et al. (2002). Lignocellulosic biomass to ethanol process design and economics utilizing co-current dilute acid prehydrolysis and enzymatic hydrolysis for corn stover. Technical Report. National Renewable Energy Laboratory (NREL), Golden (CO).
10. Wilkes, J. S., M. J. Zaworotko (1992). Air and water stable 1-ethyl-3methylimidazolium based ionic liquids. Journal of the Chemical Society, Chemical Communications, 965-967.
11. Brandt, A., J. Gräsvik, J. P. Hallett, T. Welton (2013). Deconstruction of lignocellulosic biomass with ionic liquids. Green Chemistry, 15, 550-583.
12. Kralisch, D., D. Reinhardt, G. Kreisel (2007). Implementing objectives of sustainability into ionic liquids research development. Green Chemistry, 9, 1308-1318.
13. International Organisation for Standardisation (1997). ISO 14040: Environmental Management - Life Cycle Assessment - Principles and Framework. International Organisation for Standardisation: (ISO), Geneva.

(444c) Optimal Simultaneous Production of Bio-Ibutene and Bioethanol From Switchgrass

Martin, M., University of Salamanca
Grossmann, I. E., Carnegie Mellon University
Originally presented on: 11/6/2013 9:20:00 - 9:45:00

Ibutene is an important intermediate that has been typically produced from the cracking of the C4 fraction of crude oil. It is the basis for the production of a common additive to gasoline in search for a cleaner burning fuel, the methyl-tert-butyl ether (MTBE), as well as a monomer for the polymerization such as butyl rubber. Its importance can be somehow reflected in its price, around $2/kg. However, the dependency on the crude oil and the already limited availability due to the large number of applications has increased the need for new sources of this chemical. Ibutene has recently been used as intermediate for the production of oxygenated diesel substitutes from glycerol (Chen et al 2011, Vlad et al 2010, Martín & Grossmann 2013). In this application the main drawbacks are, on the one hand, the production of diesel substitutes using a fossil fuel based intermediate and, on the other hand, the economics since from the energy and water consumption standpoint its use for the enhanced production of diesel substitutives is competitive with the process that directly sells the glycerol as byproduct.

In this work, we propose the optimal flowsheet for the production of ibutene from switchgrass. A superstructure embedding a number of alternatives is proposed. Two technologies are considered for switchgrass pretreatment, dilute acid and ammonia fibre explosion (AFEX) so that the structure of the grass is broken down. Surface response models are developed for predicting the effect of the operating conditions on the yield of each pretreatment. Next, enzymatic hydrolysis follows any of the pretreaments to obtain fermentable sugars, mainly xylose and glucose. Ibutene is obtained by fermentation of the sugars. However xylose cannot be easily converted, and thus we also evaluate the possibility of using it to produce bioethanol. Next the ibutene has to be separated mainly from CO_2 for which, PSA or membrane separation are considered while in case ethanol is produced, a multi-effect distillation column followed by molecular sieves are used to dehydrate the ethanol. The problem is formulated as an MINLP with simultaneous optimization and heat integration based on Duran and Grossmann's model (1986). Finally, an economic evaluation is performed.

The most promising process involves the use of dilute acid pretreatment and membrane purification of the ibutene. The production of ibutene using renewable sources is feasible, and may be competitive for its use to generate diesel substitutes for glycerol etherification. However, the decision related to the production of ibutene alone or the simultaneous production of ibutene and ethanol depends on the prices for bioethanol and for switchgrass. For low and current biomass cost the production of ethanol ad ibutene is interesting for bioethanol process above $1/gal.

References:

Chen, P., Min, M., Chen, Y., Wang, L., Li, Y., Chen, Q., Wang, C., Wan, Y., Wang, X., Cheng, Y., Deng, S., Hennessy, K., Lin, X., Liu, Y., Wang, Y., Martinez, B., Ruan, R., 2009. Review of the biological and engineering aspects of algae to fuels approach. Int. J. Agric. Biol. Eng. 2, 1–30.

Duran, M.A.; Grossmann, I.E. (1986) Simultaneous optimization and heat integration of chemical processes. *AIChE, J.,* 32, 123-138

Martín, M., Grossmann, I.E. (2013) Simultaneous dynamic optimization and heat integration for the co-production of diesel substitutes: Biodiesel (FAME & FAEE) and glycerol ethers from algae oil. To be Submitted

Vlad, E., Bildea, C.S. and Bozga, G., 2010, Integrated design and control of glycerol etherification processes, Bull. Inst. Pol. Iasi, LVI (LX), 4, 139-148.

(444d) Synthesis and Design of Biorefinery Processing Networks With Uncertainty and Sustainability Analysis

Cheali, P., Technical University of Denmark
Gernaey, K. V., Technical University of Denmark (DTU)
Sin, G., Technical University of Denmark
Originally presented on: 11/6/2013 9:45:00 - 10:10:00

Synthesis and Design of Biorefinery Processing Networks with Uncertainty and Sustainability analysis

Peam Cheali, Krist V. Gernaey and Gürkan Sin

Department of Chemical and Biochemical Engineering, Søltofts Plads, Building 229, Technical University of Denmark, DK-2800 Lyngby, Denmark

Chemical industries usually rely on fossil based feedstock, which is a limited resource. In view of increasing energy demands and the negative environmental and climate effects related to the use of fossil based fuels, this motivates the development of new and more sustainable technologies for processing renewable feedstocks, with the aim of bridging the gap for fuel, chemical and material production. This project is focusing on biorefinery network design, in particular for early stage design and development studies.

Optimal biorefinery design is a challenging problem. It is a multi-objective decision-making problem not only with respect to technical and economic feasibility but also with respect to environmental impacts, sustainability constraints and limited availability & uncertainties of input data at the early design stage. It is therefore useful to develop a systematic methodology to support the evaluation of processes and the generation of sustainable alternatives for identifying the optimal processing routes. One particular challenge here is to include proactively sustainability analysis as part of the synthesis of biorefinery networks. Another challenge is the handling of several sources of uncertainties such as availability and composition of renewable feedstock, the technical performance of alternative processing technologies and the availability of future markets for biorefinery products, among others.

As part of earlier work in our research group, a systematic methodology to identify/generate optimal biorefineries was developed using the superstructure-based approach, and was implemented in a computer-aided framework. The methodology consists of tools and methods including databases, models, superstructure, and solution strategies to represent, describe and evaluate various combinations of processing networks. The optimization of the network is formulated as a mixed integer nonlinear programming type of problem and solved in GAMS. The methodology was applied for designing optimal biorefinery networks considering biochemical routes. Furthermore, the methodology has also been applied to soybean oil processing networks and industrial wastewater treatment networks. In the work presented here, the methodology for designing optimal biorefinery networks was expanded with inclusion of thermochemical routes, which means the scope and the size of the biorefinery network problem was extended at the level of the database, the models and the superstructure relevant for thermo-chemical conversion routes (the conversion of biomass feedstocks (corn stover, poplar wood) to fuels and chemicals (FT-gasoline, FT-diesel, bioethanol and higher alcohols) via thermal decomposition processes like pyrolysis and gasification).,

In this study, we extend the methodology, models and database by incorporating uncertainty and sustainability analysis as well. Appropriate ranges for uncertain parameters are identified with their correlation/covariance structure and latin hypercube sampling (LHS) is used to sample parameters from their respective domain of uncertainty. The parameter samples are then used as input for solving a deterministic and stochastic optimization problem. The sustainability analysis was performed in two ways: First, it was performed retrospectively to the optimal biorefinery network solution obtained after the MINLP by using an in-house software (SustainPRO) that employs ICHEME sustainability metrics. Secondly, the sustainability analysis was included proactively as part of the MINLP optimization problem that is performed to find the optimal processing path with respect to multi-criteria assessment including technical, economics and sustainability.

The expanded database and superstructure with uncertainty and sustainability analysis form a powerful process synthesis toolbox to be used in design of future biorefineries with multi-criteria evaluation (technical and economic feasibility, environmental impact and sustainability).

(444e) Ecosystems As Unit Operations for Innovative and Sustainable Process Design

Bakshi, B. R., The Ohio State University
Urban, R. A., The Ohio State University
Ziv, G.,

Originally presented on: 11/6/2013 10:10:00 - 10:35:00

Since their development in the 1920s, the concept of unit operations has played an important role in the design of chemical processes through the development of general models and software tools. With increasing importance of reducing the environmental impact of chemical processes, the concept of unit operations expanded to include processes for reducing pollutants in the streams leaving the process boundary. Systems engineering methods based on pinch analysis and optimization have also played an important role in enhancing the efficiency of manufacturing systems. For most chemical processes, it is not practical or economically feasible to completely eliminate material emissions or waste, and the chemical industry continues to be a major producer of hazardous waste and air pollutants, including carbon dioxide.

Ecosystems have the natural ability to treat a large variety of pollutants, and provide goods and services to manufacturing systems. For example, wetlands can provide fresh water by removing contamination by chemicals such as pesticides, oils, phenolic compounds, nitrates, etc. Trees provide water, biomass and oxygen, along with the ability to remove air pollutants such as carbon monoxide, particulate matter, sulfur oxides, carbon dioxide, etc. In fact, an unsustainable system may be defined as a system that produces waste in excess of the natural cleaning capacity of ecosystems. Conversely, sustainability requires that engineering be done within ecological constraints. Unfortunately, most current sustainable engineering methods, including life cycle assessment are not able to satisfy this requirement.

In this presentation, we will "rediscover" the bounty of nature and its importance in sustaining chemical processes. This will involve expanding the concept of unit operations by including ecological processes, to enable engineering design within ecological constraints. Process flowsheets will be designed as networks of technological and ecological systems. Monetary and environmental objectives will be considered, and results for objectives such as maximum profit, zero water, and zero carbon dioxide will be compared. Application to a biodiesel manufacturing process will show that the techno-ecological network is superior to a conventional technology-only process in terms of economic and environmental objectives. Case studies of other chemical processes will demonstrate the need to modify the chemistry or processing methods to permit full benefit of ecosystems. This can lead to new innovation that may not be discovered without including ecosystems among the design options. This approach of techno-ecological networks may be extended to larger spatial scales to result in unique ways of designing and operating supply chains and entire life cycles so that they stay within ecological constraints.

(444f) Process Analysis of Continuous Catalytic Gasification As a Waste to Energy Alternative

Tillie, C. F., Cleveland State University
Riedthaler, M. J., Cleveland State University
Adhikari, S., Cleveland State University
Reeves, S. A., Cleveland State University
Gatica, J. E., Cleveland State University
Originally presented on: 11/6/2013 10:35:00 - 11:00:00

Presently, space exploration waste is either carried onboard for the duration of the mission or discarded into a trash module which is sent back into Earth's atmosphere for destruction. On the other hand, with an estimated cost of $10,000 per pound, the cost of delivering mass to locations beyond low earth orbit (LEO) locations is prohibitive.

Since approximately 15 pounds of waste are produced daily by a crew of four during a mission, the intrinsic value of this waste can be greatly enhanced by waste to energy technologies. Moreover, conceptualizing waste gasification as an in-situ alternative for processing of trash is also an efficient route to control waste, while maintaining a healthy habitable environment, during long-duration missions (Santiago-Maldonado et al., 2010).

The gasification of long-chain polymers is hypothesized to follow a mechanism of four combined reactions —two parallel oxidation reactions which are kinetically controlled, accompanied by two series reaction where oxidation products (carbon dioxide and carbon monoxide) are converted into methane and hydrogen. Integrating this process into spaceflight missions would greatly reduce the onboard mass requirements, as these products can be used as propellant fuel. Further, under certain conditions the process could be also be adapted to produce oxygen and water, two essential life support species for space missions.

This paper explores scale-up alternatives to implement low-temperature continuous gasification processes as potential waste management processes. A kinetic model formulated from batch gasification experiments is integrated into a process simulator along with a detailed thermodynamic description.

Emphasis is placed on the preparatory treatment of the waste slurries, the reactor module, and equilibrium stages for recycling reactants and recovering the catalyst. Detailed characterization of low and mid fidelity simulants is presented.

(445a) Ethyl Acetate Synthesis Using Semi-Batch Reactive Distillation

Kao, Y. L., National Taiwan University
Hsu, L. W.,
Ward, J. D., National Taiwan University
Originally presented on: 11/6/2013 8:30:00 - 8:51:00

Batch reactive distillation (BRD) is an integrated process which combines the advantages of reactive distillation and the flexibility of batch processes. Semi-batch reactive distillation (SBRD) is a design alternative for BRD in which one reactant is fed continuously into the column during the batch. Esterifications of low molecular weight alcohols (methanol–amyl alcohol) with acetic acid have been studied as model systems for the design and optimization of BRD and SBRD processes.

For such esterification processes the lowest-boiling stationary point is an azeotrope. If the azetrope is located in a two-liquid region and the aqueous phase is nearly pure water, (*n*-butanol and amyl alcohol), a BRD process is feasible because water (a product) can be removed from the column by a decanter.[1] Furthermore, Qi and Malone[2] showed that for the more difficult isopropyl acetate system in which the ternary azeotrope lies just inside the liquid-liquid zone, semi-batch reactive distillation (SBRD) operation can improve the production efficiency. Using acetic acid as an entrainer and continuously feeding it into the column during the batch can reduce the concentration of isopropyl alcohol in the distillate so that the distillate composition is closer to the water-acetate edge. Therefore the concentration of water in the aqueous phase is higher.

In this work, the more challenging ethyl acetate process is investigated. Unlike the isopropyl acetate process, BRD with a decanter is not feasible for the ethyl acetate process because the minimum-boiling ternary azeotrope is not located in the two-liquid region. To overcome this problem, researchers have considered the use of a middle-vessel column[3] and a two-column pressure swing BRD process.[4] In this work, we consider another possibility, which is the application of SBRD. Similar to the isopropyl acetate process, acetic acid is fed continuously near the top of the column to reduce the ethanol concentration in the distillate and move the distillate composition into the two liquid-liquid region. After the liquid split, high purity water can removed from the column, and the organic phase containing mainly ethyl acetate and water is collected. High purity of ethyl acetate (99 mol %) can be obtained from the separation of the collected organic product using a batch stripper with a decanter. The two step process is discussed in this work, and some important design variables including the number of reactive trays, initial amount of acetic acid added, side feed policy, reflux ratio policy, and reboiler heating policy are studied to improve the production efficiency.

References

1. Venimadhavan G, Malone MF, Doherty MF. A novel distillate policy for batch reactive distillation with application to the production of butyl acetate. *Ind Eng Chem Res.* Mar 1999;38(3):714-722.
2. Qi W, Malone MF. Semibatch Reactive Distillation for Isopropyl Acetate Synthesis. *Ind Eng Chem Res.* Feb 2 2011;50(3):1272-1277.
3. Steger C, Lukacs T, Rev E, Meyer M, Lelkes Z. A Generic Feasibility Study of Batch Reactive Distillation in Hybrid Configurations. *Aiche Journal.* May 2009;55(5):1185-1199.
4. Modla G. Reactive pressure swing batch distillation by a new double column system. *Computers & Chemical Engineering.* Nov 15 2011;35(11):2401-2410.

(445b) Coupled Permsmbr - Process Design and Development for 1,1-Dibutoxyethane Production

Pereira, C. S., Laboratory of Separation and Reaction Engineering, University of Porto - Faculty of Engineering, Porto, Portugal
Silva, V. T., LSRE - Laboratory of Separation and Reaction Engineering - Faculty of Engineering - University of Porto
Rodrigues, A. E., University of Porto - Faculty of Engineering
Originally presented on: 11/6/2013 8:51:00 - 9:12:00

The PermSMBR is an innovative technology that consists in the integration of permselective membranes with the Simulated Moving Bed Reactor (SMBR)[1]. It was already evaluated for the production of oxygenated compounds, like acetals (1,1-diethoxyethane and 1,1-dibutoxyethane) and a green solvent (ethyl lactate). The PermSMBR shows a productivity enhancement accompanied by a significant reduction on the solvent consumption when compared with other intensified technologies, as reactive distillation and/or SMBR, proving PermSMBR high potential[2-3]. However, in those studies an integrated PermSMBR was considered, which comprises a set of tubular membranes packed with a solid that acts as catalyst and selective adsorbent to water (by-product). In this work, a new configuration is presented, the coupled PermSMBR, where the tubular membranes are located after fixed-bed columns packed with the catalyst/adsorbent. By this way the membranes will not be in contact with the solid, which from an industrial point of view will be easier to implement since the process of membranes installation/replacement and clean-up is simpler. The 1,1-dibutoxyethane production is used, as an example, and each set-up features are discussed. For the studied conditions, the highest productivity achieved using the coupled PermSMBR is 73.44 $Kg_{DBE}.L_{resin}{}^{-1}.day^{-1}$ with a desorbent consumption of 0.89 $L_{n\text{-}Butanol}/Kg_{DBE}$. The results obtained indicate the coupled PermSMBR suitability for the sustainable 1,1-dibutoxyethane production.

References:

(1) Silva, V. M. T. M.; Pereira, C. S. M.; Rodrigues, A. E. Simulated Moving Bed Membrane Reactor, new hybrid separation process and used thereof. PT Patent 104496; WO Patent 2010/116335, 2009.

(2) Silva, V. M. T. M.; Pereira, C. S. M.; Rodrigues, A. E. PermSMBR—A new hybrid technology: Application on green solvent and biofuel production. *AIChE Journal* **2010,** *57*, 1840-1851.

(3) Pereira, C. S. M.; Silva, V. M. T. M.; Rodrigues, A. E. Green Fuel Production Using the PermSMBR Technology. *Ind. Eng. Chem. Res.* **2011,** *51*, 8928–8938.

Acknowledgement: The research leading to these results has received funding from the European Union Seventh Framework Programme (FP7/2007-2013) under grant agreement n° 241718, EuroBioRef.

(445c) 2-Acetoxy Propionic Acid Using Reactive Distillation - An Intermediate in Synthesis of Bio-Acrylic Acid From Lactic Acid

Kolah, A. K., Michigan State University
Lira, C. T., Michigan State University
Miller, D. J., Michigan State University
Originally presented on: 11/6/2013 9:12:00 - 9:33:00

Acrylic acid, its esters, and acrylate polymers are high volume chemicals and critical chemical building blocks used in a wide gamut of applications including superabsorbents, hard plastics, paints, detergents, and adhesives. The global acrylic acid market is estimated at 4.5 million tons and $12.4 billion annually in 2012, with a growth rate of about 4% per year.

Acrylic acid synthesized from biobased lactic acid is a drop-in replacement for the petroleum-based product synthesized from propylene. Lactic acid is a mature biobased feed stock with a proven track record of high yield (>95%) from glucose at a low cost.

Synthesis of biobased acrylic acid from lactic acid involves a two-step process. In the first step, lactic acid is reacted with acetic acid to form 2-acetoxy propionic acid (2-APA). In the second step of the process, 2-APA is cracked under high temperature to form acrylic acid. The first step reaction is thermodynamically unfavorable, with an equilibrium constant of 0.25, and results in poor conversions if carried out in a batch reactor even with high mole ratios of acetic acid to lactic acid.

In this work, the selective formation of 2-APA by reaction of lactic acid with acetic acid has been carried out using continuous reactive distillation. The reaction takes place in a six-meter tall, 50 mm ID Pyrex column operated at atmospheric pressure, and is catalyzed by Amberlyst-15 cation exchange resin supported inside Sulzer Katapak-SP11 structured packing. Lactic acid feed stocks from 20 wt% to 85 wt% in water were used in experiments along with glacial acetic acid. High conversion of lactic acid and its oligomers were observed (>98%). When 20% lactic acid was used, a selectivity of 96% to 2-APA was obtained.

An unusual observation in the experimental work was the enhancement of the acetic acid – water separation by the presence of lactic acid. Because of this enhancement, it is possible to obtain a distillate stream containing 99% water from the reactive distillation column while retaining acetic acid within the column where reaction occurs.Ternary VLE experiments were performed to obtain the NRTL parameters describing this system.

Experimental studies are complemented by process analysis using AspenPlus simulation software. By implementing a pseudo-homogeneous kinetic model generated in our laboratories from batch experimental kinetic data, and descsribing vapor-liquid equilibria based on experimental data, we are able to predict the experimentally observed pilot-scale column behavior.

It can be concluded that 2-APA can be synthesized very efficiently from aqueous lactic acid solutions by reaction with acetic acid using a continuous reactive distillation column.

(445d) Novel Technology for Producing High Purity n-Hexane and Benzene

Yu, Y., Tianjin University
Li, Y., Tianjin University
Geng, Z., Tianjin University
Ma, J., Tianjin University
Zhang, Z., Tianjin University
Zhang, M., Tianjin University
Originally presented on: 11/6/2013 9:33:00 - 9:54:00

N-hexane is one of the most representative non-polar solvent. It is widely used in pharmaceutical, chemical, polymer materials etc., and owning a great market demand. Benzene is one of the most important petrochemical materials. The yield and production technology of benzene stand for the petrochemical development level of the country. However, the traditional separation and purification technologies, for example, molecular sieve adsorption, batch azeotropic distillation and extractive distillation etc., are difficult to get high yield production of n-hexane and benzene with high purity. Moreover, the energy consumption is high in these processes. Our team has developed a series of new technologies coupling the normal distillation and extractive distillation to produce high purity n-hexane and benzene, and iso-hexane as a by-product. N-methyl pyrrolidone was chosen as the extractant in the process. C3~C8 alkane mixtures from raffinate of aromatics extraction unit, pyrolysis gasoline or straight-run gasoline were used as feed. Gradual separation was used to get the final products. Firstly, the iso-hexane was obtained after removal of the impurities with boiling point lower than 2-methylpentane. Then, the remaining materials were separated by extractive distillation to acquire n-hexane product. Finally, the benzene product was got by further separation. The multi-effect thermal coupling technique was employed in the distillation unit. With the heat integration and optimization, the usage of steam and cooling water was saved. The usage of extractant was also reduced by the way of the cycling between the extractive distillation column and the recovery column. To prevent impurities accumulated in the extractant, removal of the heavy component impurities was implemented by taking a small part of material into the falling film evaporation system. High purity n-hexane and benzene product were finally produced by the above mentioned method. The purity of n-hexane product is higher than 99 wt% with the content of benzene is lower than 100 ppm. At the same time, the concentration of benzene product is higher than 99.9 wt%. Currently, a 30,000 MTA n-hexane and benzene plant has been built up using this technology.

(445e) Online Monitoring of Emulsion Polymerisations Using Raman Spectroscopy

Houben, C., University of Cambridge
Nurumbetov, G., University of Warwick
Haddleton, D.,
Lapkin, A., University of Cambridge
Originally presented on: 11/6/2013 9:54:00 - 10:15:00

Many polymer products are manufactured using batch and semi-batch reactors. Typically the process parameters are fixed and are not varied during the reaction. Unexpected changes in operating conditions can lead to variations in polymer structure and purity, resulting in significant losses due to off-spec materials. An approach addressing the complex issues of real-time process control of large-scale emulsion polymerisation processes is to use advanced models and on-line sensors. A specific objective is to implement a model-based predictive control based on Raman spectroscopy.

(445f) Process Intensification in Millistructured Extraction Columns

Holbach, A.,
Caliskan, E., TU Dortmund University
Kockmann, N., TU Dortmund University
Originally presented on: 11/6/2013 10:15:00 - 10:36:00

Process intensification in millistructured extraction columns

The downstream process of biochemical and pharmaceutical products can account up to 80% of the cost of production. [1,2] To reduce the cost it is necessary to develop new process intensified equipment with higher time-volume-yield. Process intensification is one important part to make the industry more economical and ecological. Some mass or heat transport limited chemical reactions are only technical possible or cost-efficient with process intensified equipment. The small channels and structures in this equipment gives the potential to control the fluids on the sub-millimeter scale. The known characteristics of transport phenomena enable an easy and fast scale up for bulk chemicals from lab to industrial scale.

We investigated the process intensification in millistructured extraction columns with an inner diameter of 15 mm, see figure 1. The column has 20 stirred cells and an active extraction high of 440 mm. The total internal volume with head and bottom is 220 ml with total height of 760 mm. With stirred cells we achieved high energy input, high specific surface between the two phases and thus good mass transfer.Â The backflow was decreased with narrow plates between the plates. Thus, the dispersed phase coalesces at the plates and plugs the stirred cells. With axial pulsation of the fluid the coalescence area is destroyed and the dispersed phase is transported into the next cell. This new technique allows for controlled flow of the dispersed phase into the next stage. The retention times for counter current processes can be adjusted for transport limited processes or reactive extraction systems.

The hydrodynamics and mass transfer inside the column was characterized with the European Federated Chemical Engineering (EFCE) testsystem n-butylacetate(d)/acetone/water(c). The mass transfer from continuous-to-disperse (c-d) and from dispers-to-continuous (d-c) and the droplet size distribution with extraction efficiency was investigated. Different to conventional columns, the energy input inside the stirred cells are higher and the influence on the coalescenc and mass transport direction is not so large. We achieved specific surfaces between the two fluids up to 3200 m^2/m^3 and measured the mass transfer up to 7 theoretical extraction stages per active high (H = 440 mm). This correspondsto 16 theoretical stages per meter, see figure 1. The experiments show that higher total throughput yields higher extraction efficiencies, hence, high mass flow rates are preferred.

The enantioselective extraction of (R/S)-α-cyclohexyl-mandelic acid will be investigated as an example; see the molecules in figure 2. The enantioselective extraction is an organic/aqueous reactive extraction. The enantiomers create in the aqueous phase a complex with hydroxy-propyl-ÃŸ-cyclodextrin. The (S)-complex is preferred to the (R)-complex, so that the extraction is selective. The whole process is mass transport limited and has a low selectivity. In conventional columns the extraction is impossible, because the retention time of the dispersed phase and the specific surface between the two fluids are too small. In process intensified extraction columns with high specific surfaces and many extraction stages the counter current extraction is possible. First results show a considerable separation of the two enantiomers.

[1] Lowe C. R., Lowe A. R., Gupta G., New developments in affinity chromatography with potential application in the production of biopharmaceuticals, J. Biochem. Biophys. Methods 49, 2001, 561?574

[2] Strube J., Sommerfeld S., Challenges in biotechnology production?generic processes and process optimization for monoclonal antibodies, Chemical Engineering and Processing, 44, (10), 2005, 1123-1137

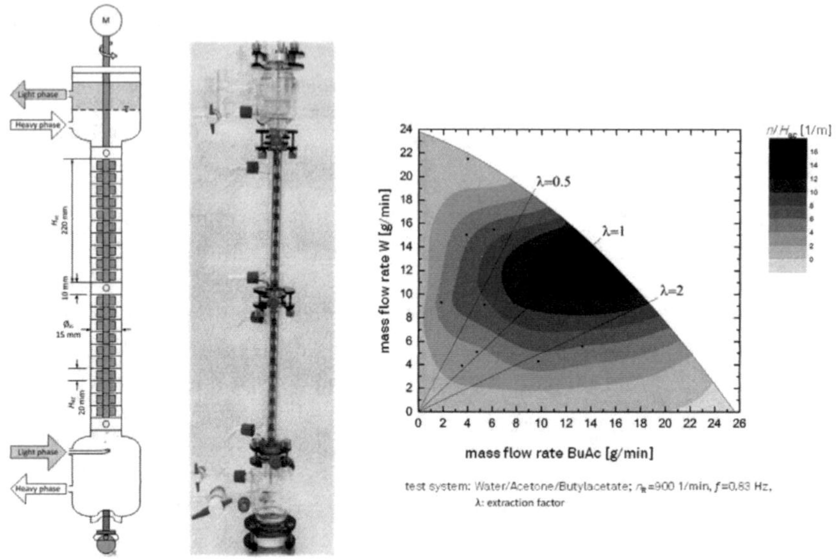

Figure 1: Millistructured extraction column; left side: sketch and picture of the column with two sections; right side: experimental extraction efficiency with different mass flow rates

(S)-α-cyclohexyl-madelic acid

(R)-α-cyclohexyl-madelic acid

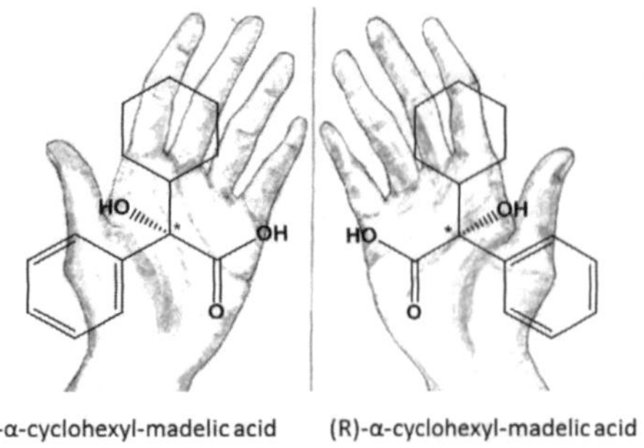

Figure 2: Molecular structure of (S) and (R)- α-cyclohexyl-madelic acid

(445g) A NOVEL Comparison Among Intensified and Traditional Complete Ethyl Acetate Productions Designs

Santaella, M., Universidad Nacional de Colombia
Narvaez, P. C., Universidad Nacional de Colombia
Orjuela, A., Universidad Nacional de Colombia
Originally presented on: 11/6/2013 10:36:00 - 10:57:00

Over the last decade different proposals of process intensification have been made for the improvement of the esterification reaction between acetic acid and ethanol. The dissimilar considerations and in some cases unrealistic assumptions made by the authors during modelling limit the possibility to compare among different production processes. In addition to this, different inlet compositions and different product specifications represent a major barrier to ensure an objective comparison between different processing technologies.

In this paper, the conventional production process, the reactive distillation (RD) and the reactive dividing wall column (RDWCD) process for the production of urethane grade ethyl acetate (99,5%) have been simulated using a commercial software (Aspen One 7.2). Glacial acetic acid and azeotropic ethanol were used as raw materials, and different catalysts were evaluated (self-catalytic, sulphuric acid, Amberlyst 15, Amberlyst 70). A validated NRTL-HOC thermodynamic model was used to describe phase equilibria. Chemical kinetic models were obtained from literature. A preliminary evaluation of reactive residue curve maps obtained with different catalysts and residence time ranges between 0-500min allowed to compare feasible column profiles.

Capital costs and energy savings for reactive distillation are predicted between 15-80% over classical configurations of a reactor followed by separation units. Dividing wall column, Petluk column and similar thermally coupled configurations predict 30-50% energy savings over conventional two distillation column processes. The RDWCD also predict 30-50% energy reduction over simple RD.

(470a) A Universal Procedure for Crude Glycerol Purification From Different Feedstocks in Biodiesel Production: Experimental and Simulation Study

Xiao, Y., Purdue Univeristy
Xiao, G., Southeast University
Varma, A., Purdue University
Originally presented on: 11/6/2013 12:30:00 - 12:52:00

With increasing supply and decreasing price, from the economic point of view, it is important to utilize crude glycerol, the main by-product of biodiesel production, to manufacture high value-added chemicals. Since crude glycerol typically contains less than 65 wt % glycerol and a variety of other components, its purification is the first step for its utilization. Owing to the wide variety of triglycerides, alcohols, catalysts and separation processes used in biodiesel production, the composition of crude glycerol from different biodiesel production methods varies widely. For this reason, glycerol purification processes are also generally different for the various crude glycerol feedstocks. In the present work, we have developed a universal procedure for crude glycerol purification, and utilized it for crude glycerol samples from two biodiesel production companies, experimentally upgrading both to >94 wt % purity. On Aspen Plus software platform, the purification procedure was simulated using a process model based on modeling of phase equilibrium behavior and modeling of NaCl solubility in ethanol-glycerol mixtures, to obtain good match with the experiments. Using the same model, two other glycerol purification processes described in the literature were also simulated and high purities were also obtained according to the simulation results. We thus demonstrate that the developed universal procedure is suitable for purification of crude glycerol obtained from different biodiesel production technologies.

(470b) Two-Stage in Situ Gas Stripping for Enhanced Butanol Fermentation in a Fibrous Bed Bioreactor and Energy-Saving Product Recovery

Xue, C.,
Yang, S. T., The Ohio State University
Originally presented on: 11/6/2013 12:52:00 - 13:14:00

Acetone-butanol-ethanol (ABE) fermentation with a hyper-butanol producing *C. acetobutylicum* JB200 was studied for its potential to produce a high titer of butanol that can be readily recovered with two-stage gas stripping. In batch fermentation without gas stripping, 80.0 g/L glucose was consumed by *Clostridium acetobutylicum* JB200 immobilized in a fibrous bed bioreactor, producing 7.9 g/L acetone, 16.2 g/L butanol and 1.4 g/L ethanol. When the two-stage gas stripping was incorporated into fed-batch fermentation, 49.2 g/L acetone, 113.3 g/L butanol and 9.7 g/L ethanol were produced from 474.9 g/L glucose in six feeing cycles over 326 h, and butanol yield and productivity increased by 20% and 16.7%, respectively. The higher productivity was attributed to the reduced butanol concentration in the fermentation broth by gas stripping that alleviated butanol inhibition to cells, whereas the increased butanol yield could be attributed to the reduced acids accumulation as most acids produced in acidogenesis were reassimilated by cells for ABE production. Meanwhile, the recovered condensate containing 150.5 g/L butanol and 195.9 g/L ABE was collected from the first-stage gas stripping. Since butanol has the solubility of 7.7% (w/w) in water, the condensate obtained has two phases due to the butanol concentration in the condensate far exceeding its solubility in water. Consequently, after phase separation, the organic phase contained ~610 g/L butanol and ~660 g/L ABE and the aqueous phase contained ~80 g/L butanol and ~120 g/L ABE, which was then subjected to second-stage gas stripping. The mixture of the organic phase and condensate of the second-stage gas stripping contained ~430 g/L butanol and ~560 g/L ABE, which could be purified at much less energy consumption. Compared to conventional ABE fermentation, the fed-batch fermentation with two-stage gas stripping has the potential to reduce energy consumption and water usage in n-butanol production from glucose.

(470c) Improvement of Biofuel Production By Fungal Pretreatment and Periodic Flushing System

Yao, W.,
Nokes, S. E., University of Kentucky
Originally presented on: 11/6/2013 13:14:00 - 13:36:00

In this study we investigated the fermentation performance (quantified by solvent and acid production) of fungal pretreated (*Phanerochaete chrysosporium)* agricultural residues (corn stover or miscanthus). The fermentation process employs sequential cultivation of the cellulolytic bacterium *Clostridium thermocellum (C. thermocellum)* at 65 °C followed by the cultivation of the butanol-producing strain, *Clostridium beijerinckii* (*C. beijerinckii*) at 35 °C. Pretreated agricultural residues were fermented in a glass column fitted with porous polyethylene discs at the base of the column, allowing periodical flushing of the column with culture medium while retaining the solid substrate. The hypothesis of this study is that fungal pretreatment improves cellulose hydrolysis and the repeated flushing process helps relieve the metabolic inhibition which occurs in high- solid substrate fermentation. Time profiles of soluble sugars, solvents and acids were quantified during the entire culture process. A parallel study quantifying the efficacy of reinoculation of *C. thermocellum* or *C. beijerinckii* on product generation was performed to optimize the repeated flush process.

(470d) Solid Phase Reactive Simulated Moving Bed Chromatographic Separation System for Biofuel and Biochemical Production

Kanchanalai, P.,
Realff, M., Georgia Institute of Technology
Kawajiri, Y., Georgia Institute of Technology
Originally presented on: 11/6/2013 13:36:00 - 13:58:00

Efforts to produce new cleaner and more sustainable liquid fuels and chemicals from lignocellulosic biomass have created the incentive to develop efficient and cost-effective processes for pretreatment of this type of biomass. Hydrolysis of lignocellulosic materials for sugar production can proceed via different pretreatment and hydrolysis technologies. Two well-known classes of hydrolysis technology are enzymatic and acid hydrolysis. Hydrolysis by both diluted and concentrated acid media has been reported having a faster rate of reaction compared to the enzymatic hydrolysis pathway[1], but has drawbacks both in terms of byproduct formation and the cost of disposal of the acid streams after hydrolysis. In the diluted acid hydrolysis process, a high operating temperature is required to break down the structure of the material. These severe conditions cause sugar decomposition and generate side products such as furfural, HMF, acetic acid, and levulinic acid which can inhibit the fermentation process for bioethanol production raising overall capital and operating costs. In the concentrated acid hydrolysis process, a smaller amount of sugar degradation is reported due to lower operating temperatures. Nevertheless, the main drawbacks are the expensive process equipment to prevent corrosion and a significant cost of acid recovery.[2]The synthesis of a new process for acid hydrolysis that avoids these two drawbacks is our research goal.

Considerable effort has been made to improve both the reactor designs and the separation process to achieve higher sugar yield and productivity. Several reactor designs have been proposed for acid hydrolysis of lignocellulosic materials such as batch, plug flow, and percolation reactors. A percolation reactor is a packed-bed-flow reactor where the sugar is removed as it formed; therefore, sugar concentration increases with smaller byproduct formation compared to the batch and plug flow reactors.[3] Another reactor configuration is a counter-current reactor with a reverse flow between a biomass and a liquid stream was investigated and found to have a significant improvement of sugar yield due to the shorter time of sugar decomposition.[4] However, the movement of a solid stream in this process increases the complexity of the design and operation of the reactor. Therefore, a progressing batch reactor (PBR) design has been proposed to mimic the countercurrent flow of diluted acid and biomass.[5] The key principle of this reactor configuration is the movement of the feed location of the acid to different percolation reactors arranged in series. The main advantage of this configuration is the realization of countercurrent scheme while maintaining its simplicity of percolation reactor operation. For the separation process, simulated moving bed (SMB) chromatography has gained interest to separate the acid from sugars in the hydrolyzate.[6,7] The underlying principle of the SMB, similar to the PBR, is that the fluid is in contacted with the solid phase where the feed and outlet streams switch to simulate the movement of solid phase to achieve a countercurrent flows.[8]

The disadvantages of the acid hydrolysis reaction motivate this work to develop reactive separation in which reaction and separation are integrated to avoid byproduct formation, improve sugar yield, and reduce recycle cost. Previous studies have explored the reactive separation principle with the simulated moving bed chromatography which has been applied to both liquid gas phase reactions.[9]However, no previous work has been considered a combination of SMB and reactor where the reactants are in the solid and liquid phase.

This work proposes a new reactive separation process system termed solid-phase reactive separation system (SPRSS) which integrates the PBR and SMB units and aims to apply this to the acid hydrolysis of lignocellulosic materials for sugar production before fermentation to produce bioethanol. Both PBR and SMB have similar principles of the movement of the liquid feed to imitate the countercurrent movement of solid and liquid phase. The overarching goal is to improve the sugar yield and sugar product concentration while minimizing the sugar decomposition reaction and undesired product formation. Moreover, the acid consumption is expected to be further reduced and therefore decrease the cost of acid recycle. Optimization strategy using a superstructure formulation will be applied to find the optimal process parameters and process configurations of SPRSS to maximize both of the sugar concentration and the sugar yield while minimizing the byproduct formation.

Reference

(1) Lenihan, P.; Orozco, A.; O'Neill, E.; Ahmad, M. N. M.; Rooney, D. W.; Walker, G. M. Dilute acid hydrolysis of lignocellulosic biomass. *Chem Eng J* **2010,** *156*(2), 395-403.

(2) Taherzadeh, M. J.; Karimi, K. Acid-Based Hydrolysis Processes for Ethanol from Lignocellulosic Materials: A Review. *Bioresources* **2007,** *2*(3), 472-499.

(3) Lee, Y. Y.; Iyer, P.; Torget, R. W. Dilute-Acid Hydrolysis of Lignocellulosic Biomass. In *Recent Progress in Bioconversion of Lignocellulosics*, Tsao, G. T.; Brainard, A. P.; Bungay, H. R.; Cao, N. J.; Cen, P.; Chen, Z.; Du, J.; Foody, B.; Gong, C. S.; Hall, P.; Ho, N. W. Y.; Irwin, D. C.; Iyer, P.; Jeffries, T. W.; Ladisch, C. M.; Ladisch, M. R.; Lee, Y. Y.; Mosier, N. S.; Mühlemann, H. M.; Sedlak, M.; Shi, N. Q.; Tsao, G. T.; Tolan, J. S.; Torget, R. W.; Wilson, D. B.; Xia, L., Eds. Springer Berlin Heidelberg: 1999; Vol. 65, pp 93-115.

(4) Song, S. K.; Lee, Y. Y. Countercurrent Reactor in Acid-Catalyzed Cellulose Hydrolysis. *Chem Eng Commun* **1982,** *17*(1-6), 23-30.

(5) Wright, J. D.; Bergeron, P. W.; Werdene, P. J. Progressing Batch Hydrolysis Reactor. *Ind Eng Chem Res* **1987,** *26*(4), 699-705.

(6) Heinonen, J.; Tamminen, A.; Uusitalo, J.; Sainio, T. Ethanol production from wood via concentrated acid hydrolysis, chromatographic separation, and fermentation. *J Chem Technol Biot* **2012,** *87*(5), 689-696.

(7) Xie, Y.; Chin, C. Y.; Phelps, D. S. C.; Lee, C. H.; Lee, K. B.; Mun, S.; Wang, N. H. L. A five-zone simulated moving bed for the isolation of six sugars from biomass hydrolyzate. *Ind Eng Chem Res* **2005,** *44*(26), 9904-9920.

(8) Dunnebier, G.; Engell, S.; Klatt, K. U.; Schmidt-Traub, H.; Strube, J.; Weirich, I. Modeling of Simulated Moving Bed chromatographic processes with regard to process control design. *Comput Chem*

Eng **1998,** *22*, S855-S858.

(9) Hashimoto, K.; Adachi, S.; Noujima, H.; Ueda, Y. A New Process Combining Adsorption and Enzyme Reaction for Producing Higher-Fructose Syrup. *Biotechnol Bioeng* **1983,** *25* (10), 2371-2393.

(470e) Butanol and Hydrogen Production From Xylan By Clostridium Sp. BOH3: Simultaneous Saccharification and Fermentation

Rajagopalan, G., National University of Singapore (NUS)
He, J., National University of Singapore (NUS)
Kun-Lin, Y., National University of Singapore (NUS)
Originally presented on: 11/6/2013 13:58:00 - 14:20:00

Title

Butanol and hydrogen production from xylan by *Clostridium* sp. BOH3: simultaneous saccharification and fermentation

Abstract

Butanol and hydrogen are recognized as fossil-fuel alternatives. They can be simultaneously produced from anaerobic fermentation of simple sugars. Recently, a new strain *Clostridium* sp. BOH3 was observed to produce butanol and hydrogen from xylan. While the most of solventogenic *Clostridia* spp. require simple fermentable sugars such as glucose for the production of butanol, sp. BOH3 utilizes xylan (10 g/l) effectively because of its high level xylanase expression. In the present study, a medium is designed for the production of xylanase, butanol and hydrogen using response surface methodology. The medium is composed of two major components, xylan as carbon source and ammonium sulphate as nitrogen source. With the help of central composite design, the direct and interactive effects of xylan and ammonium sulphate on xylanase, butanol and hydrogen production were analysed. The optimized medium, which supports maximum production of xylanase, butanol and hydrogen, contained 49.4 g/l of xylan and 19.8 g/l of ammonium sulphate. It supports the production of 6.1 U/ml of xylanase, 7.6 g/l of butanol and 16.2 ml/l.h of hydrogen. Apparently, simultaneous saccharification and fermentation (SSF) of xylan by using BOH3 is possible, thanks to the expression of xylanse by BOH3. To improve this SSF process, xylanase produced by BOH3 was purified by using one-step Q-anion column chromatography and supplemented into the next batch. Every attempt of purification yields 43Â±1.5% of enzyme, and its supplementation to the next batch results the xylanase activity of 2.5Â±0.1 U/ml at the beginning of fermentation. Due to this addition of xylanase, production of butanol and hydrogen is increased to 16.05Â±0.5 g/l and 28.3Â±0.8 ml/l.h respectively. The butanol and hydrogen yields are increased to 0.33Â±0.01 g/g and 0.58Â±0.008 ml/l.h/g respectively. These results suggest that improved SSF process enhances the butanol and hydrogen yield about 2 fold. This approach would also facilitate the effective recycling of potential enzyme produced in the fermentation process.

(470f) An Integrated Approach for the Production of Bioethanol and High Value Bio-Products From Lignocellulosic Biomass

Karapatsia, A.,
Penloglou, G.,
Pappas, I.,
Kiparissides, C., The Petroleum Institute
Originally presented on: 11/6/2013 14:20:00 - 14:42:00

In the last centuries, people in the developed countries became almost totally dependent on fossil sources, in order to produce energy fuels and other basic chemicals required for maintaining a high life quality. However, with the depletion of fossil oil-resources, the ever increasing oil-price and the negative environmental impact due to the gaseous emissions associated with fossil fuels consumption, the world is actively searching for alternative energy sources. Plant biomass has the potential to partially satisfy the global needs in basic fuels and chemicals, while minimizing environmental impact and increasing sustainability.

Lignocellulosic materials (i.e., dedicated energy crops, such as perennial grasses, forestry co-products, agricultural residuals and vegetable waste etc.) are excellent examples of alternative carbon and energy sources with a large potential for flexible production of bio-based products. Lignocellulose, the most abundant polysaccharide-containing biomass available in the world, is an extremely complex and widely varying nano-scale composite, combining cellulose, hemi-cellulose, and lignin, along with a variable level of extractives. The presence of cellulose and hemi-cellulose, therefore, makes lignocellulosic feedstock a potential candidate for biochemical conversion processes, representing, at the same time, a promising outlet to security of supply issues for future production of bio-based products.

Among the various lignocellulosic feedstocks, the dedicated perennial herbaceous species with a C_3 photosynthesis system (e.g., *Phalaris aquatica* L. - Harding Grass, HG) represent a potential solution that can improve further the robustness and reliability of this renewable, lignocellulosic-based source. Characteristics of these types of dedicated perennial plants are their fast growth, their high biomass yields, the low input costs, the region-specificity, and the annual or, in many cases, less-frequent harvests. Moreover, they can grow on non-prime agricultural land, such as under-utilized and abandoned farmland, and do not compete for prime agricultural land required by commodity crops, grown for the food and feed markets.

The sugar monomers in lignocellulosic biomass, existing in the form of polysaccharides, are not readily available for bioconversion, due to the recalcitrance of plant cell wall structure. As a result, a multi-step process is required in order to liberate the carbohydrates for bioconversion to biofuels (e.g., ethanol), biochemicals (e.g., succinic acid), biopolymers (e.g., polyhydroxybutyrate, PHB) and other high value bio-based products. Dilute acid pretreatment followed by enzymatic hydrolysis is a typical process used to convert lignocellulosic biomass to fermentable sugars. The fermentation of both hexoses and pentoses via specific microbial factories into the desired bio-products follows the above treatment steps. Additionally, the separation of the desired products from the fermentation broth is of highly importance.

In the present work, the above-described integrated biochemical process was pursued with respect to the production of bioethanol from lignocellulosic biomass, involving three major stages, namely i) pretreatment, ii) enzymatic hydrolysis and iii) fermentation, which were studied and optimized separately: The optimization of the biomass pretreatment process was experimentally investigated following a statistical experimental design (Taguchi), via the maximization of hemi-cellulose solubilization and the recovery of pentose sugar monomers. The three principal variables studied were the reaction temperature, the concentration of sulfuric acid and the hydrolysis time. The parameters were controlled in 3 different levels (high, low and intermediate). From a series of experiments, the optimal conditions were identified. More specifically, the experimental run using 1.5% sulfuric acid at 120 °C for 45 min, resulted in approximately 80% conversion of the total available hemi-cellulose into pentose sugar monomers. Moreover, the results revealed that the combination of factors with mild conditions yield the maximum amount of pentoses, while enhancing the hydrolysis of cellulose in the following process step.

The enzymatic hydrolysis of the solid residue of the pretreatment stage was also optimized, by using a statistical experimental design (Box Denken), including the addition of surfactants. For the experimental study of the saccharification of cellulose four principal variables were selected and tested (biomass content, amount of enzyme, hydrolysis time and surfactant concentration-PEG 4000) at three different levels (high, low and intermediate). The mixture of commercial enzymes comprised cellulase (Celluclast 1.5L) and beta-glucosidase (Novozyme 188), derived from the fungi *Trichoderma reseei* and *Aspergillus niger*, respectively. From the experimental results, it was concluded that the optimal saccharification conditions were: 4% solids concentration, 20 FPU enzymes, 0.04 g/g solid surfactant concentration and a 72 h hydrolysis time. Furthermore, an empirical regression model was developed for the prediction of the maximum glucose production in the range of the selected variables. The maximum value of the concentration of glucose was equal to 12.9 g/l. Experiments were also carried out with increased amounts of initial solids content, in order to increase the final concentration of glucose in the hydrolyzate solution, a prerequisite for the subsequent fermentation stage.

The fermentation (cultivation of the yeast *Saccharomyces cerevisiae*) of the recovered glucose in the hydrolyzate solution, derived from the enzymatic hydrolysis stage, was held both in flask and bioreactor units. Initially, the optimum conditions (i.e., agitation speed, pH, inoculum size, and medium composition in terms of glucose and nitrogen sources concentrations, yeast strain) were identified under batch conditions in flask experiments. Experiments were also carried out under fed-batch mode using similar cultivation conditions. The optimal fed-batch policy identified in the flask experiments, resulted in an enhanced ethanol production: maximum ethanol concentration equal to 18.84 g/l), simultaneously with an overall productivity 0.82 g/(l·h). Subsequently, the optimal fermentation fed-batch policy was transferred to a two-liter bioreactor unit, where four different fed-batch runs were evaluated. It was found that a simple continuous feeding policy can significantly increase the bioethanol concentration and the overall productivity. The maximum ethanol productivity was equal to 1.97 g/(l·h).

Beyond the fermentative production of bioethanol from the available hexose monomers (mainly glucose), in the present research program, the synthesis of additional high value products was pursued. More specifically: i) the production of succinic acid by the fermentation of the available pentose monomers, liberated from the pretreatment stage, by *Actinobacillus succinogenes*, ii) the synthesis and recovery of biodegradable polyhydroxybutyrate (PHB) via the cultivation of *Azohydromonas lata*, and iii) the thermo-chemical production of plant nanocellulose directly from the dried biomass feedstock. These production lines could be part of an integrated biorefinery, combining the production of bioethanol with the production of high value-added materials, thus, significantly contributing to the economic viability of the biorefinery plant.

(470g) Evaluation of Ethanol Production From Renewable Cellulosic Resources Using Process Simulation Tools

Petrides, D., INTELLIGEN, INC.
Siletti, C., INTELLIGEN, INC.
Carmichael, D., Intelligen, Inc
Originally presented on: 11/6/2013 14:42:00 - 15:04:00

Over the past three decades there has been intense investigation on the development of fuel producing processes that are based on the use of renewable agricultural materials as feedstock. This activity is driven primarily by the quest for fuel self-reliance and carbon oxides emission reductions. The main effort has been concentrated on bio-ethanol and bio-diesel which have been shown to give motor engine performance similar to that of conventional petroleum based fuels. In addition to product characteristics, however, process economics play an equally important role in any successful product commercialization. In this work, realistic process simulation models have been developed in order to analyze the economics of corn-stover to ethanol conversion. The evauation indicates the the raw materials dominate the cost for this process. The process displays an economy of scale with a trade off off capital and operating costs. This poster presentation will illustrate how such models can guide R&D work and facilitate process optimization.

(478a) Enhanced Yield of Bio-Gas Production With Modified Themo-Reactor Configuration and Separation of CO2

Tahir, M. S., University of Gujrat
Saleem, M., Institute of Chemical Engineering and Technology, University of the Punjab, Lahore, Pakistan
Originally presented on: 11/6/2013 12:30:00 - 12:50:00

Pakistan is currently facing severe energy crisis with significant shortfall. A number of short-term technical solutions are available but are not adequate to solve the issue in the longer run. Production of bio-gas with modified thermo-reactor from the animal waste at regional level shows potential of reducing burden on total natural gas resource in Pakistan.

This research study investigates the effects of solar thermo-rector configuration on the overall yield and composition of bio-gas from dung. The designed capacity of the bio-gas plant is 10 m^3. The solar thermo-reactor employs internal heating system with coils using warm water heated from the solar energy. The bio-gas reactor was operated over the range of 25-55 ^{0}C to observe effects on bio-gas yield and composition. It was observed that the increase in temperature within the thermo-reactor resulted in significant increase in yield of bio-gas i.e. thermo-reactor operated at 50-55 ^{0}C (thermophilic). Operation time and plant size of state of the art process can be reduced.

The calorific value of bio-gas obtained at 50-55 ^{0}C was slightly higher than calorific value of bio-gas obtained at mesophilic condition (25-35°C) due to increased production of CH_4. Thermophilic operation of bio-gas plant also enhances the methane content in the bio-gas i.e. 57-60 % vol of methane. Bio-gas produced was purged from bottom and cold water from top a packed column and decreases in CO_2 was also recorded. The results were scaled up from pilot scale capacity to regional level in order to estimate potential of bio-gas production

(478b) Microwave-Assisted Catalytic Glycolysis of Poly(ethylene terephthalate) By Using Ionic Liquids

Liu, Y., Donghua University
Xing, Y., Donghua University
Originally presented on: 11/6/2013 12:50:00 - 13:10:00

Poly(ethylene terephthalate) (PET) was widely used for beverage bottles, electrical/electronic instruments and household wares. As a consequence of dramatically increasing consumption, recycling of post-consumer PET products has become an important environmental opportunity for sustainable usage in society. The presented study is aiming to investigate the use of ionic liquids as catalysts for the glycolysis of PET. A series of novel ionic liquids using 1-butyl-3-methylimidazolium chloride (BMIMCl) and zinc halides (ZH) as precursor with different molar ratios was first synthesized. The glycolysis of PET by microwave-assisted method using the synthesized ionic liquids as catalysts and glycol as solvent was also studied. The influences of the type of zinc halide, the molar ratio of BMIMCl:ZH, the dosage of ionic liquid and solvent, reaction time and reaction time on the yield of recovered terephthalic acid were investigated. The synthesized ionic liquids and the glycolysis product were characterized by X-ray diffractometer, infrared spectrometric analyzer, ^{1}H NMR, differential scanning calorimeter and DSC. The results showed that the synthesized ionic liquids exhibited higher catalytic activity for the glycolysis of PET by microwave-assisted method□The halogen anion has a key influence on the glycolysis of PET. A study on the recycling of the ionic liquid shows that it could be used repeatedly. A mechanism of the degradation of PET was proposed and the kinetics of the reaction was needed further investigated.

(478c) Degradation of Rhodamine-B By Heteropolyoxometalate Ionic Liquid Under Visible Light

Liu, Y., Donghua University
Chen, Y., Donghua University
Xing, Y., Donghua University
Originally presented on: 11/6/2013 13:10:00 - 13:30:00

The photo-catalytic degradation of Rhodamine-B(RhB) dye was successfully achieved using different heteropolyoxometalate phosphonium ionic liquids(PWPILs) under visble light irradiation. A series of heteropolyoxometalate phosphonium ionic liquids were first prepared from phosphotungstic acid and different quaternary phosphonium salts trihexyl(tetradecyl)phosphonium bromide, tributyl(octyl)phosphonium bromide and tetrabutyl phosphonium bromide with different molar ratioa. The melting points of PWPILs, all lower than 100°C,were characterized by differential scanning calorimeter(DSC). The results show that the characterized peaks of phosphotungstic anion in IR spectra of PWPILs are consistent with that of phosphotungstic acid. This indicates that the quaternary phosphonium cations do not change the Keggin structure of heteropoly acid anion. The influences of experimental parameters, such as the carbon chain length in phosphoniums, the number of free hydrogen in phosphotungstic anions ,addition of different catalysts, irradiation time and the concentration of RhB solution on the photo-catalytic degradation of RhB were investigated. The photo-degradation efficiency was characterized by decolourization rate of RhB. The results indicated that $[P(C_{14}H_{29}(C_6H_{13})_3]_3[PW_{12}O_{40}]$ showed the best photo-catalytic degradation efficiency.
The RhB solution (2×10-4mol/L) was almost decomposed catalyzed by $[P(C_{14}H_{29})(C_6H_{13})_3]_3[PW_{12}O_{40}]$ under UV-Vis light irradiation for 3 hours. The degradation rate was more than 95%.

(478d) Metal-Containing Ionic Liquid As An Effective Catalyst for the Degradation of Poly(ethylene terephthalate) in Ethylene Glycol

Lu, X., Institute of Process Engineering, Chinese Academy of Sciences
Wang, Q., Institute of Process Engineering, Chinese Academy of Sciences
Zhang, S., Institute of Process Engineering, Chinese Academy of Sciences
Originally presented on: 11/6/2013 13:30:00 - 13:50:00

The vast emissions of poly(ethylene terephthalate) (PET) wastes have caused serious environmental pollution and resource waste. Its effectively recycling has become a global problem which must be urgently solved.

Chemical recycling is an attractive process because PET can be degraded into monomers which can be used as raw materials to produce virgin plastic products. Glycolysis, as an important chemical recycling method, has received extensive concern due to its advantages of less volatile solvents, high conversion of PET, giving monomers and easily continuous production. But there are some drawbacks, such as high temperature, high pressure, low reaction rate and difficult separation between catalyst and product. Many researchers devoted to developing new catalysts to overcome the drawbacks above. Ionic liquids have been reported that could effectively catalyze the degradation of PET to give BHET monomers with high purity and make the reaction condition mild. But the conversion of PET and the selectivity of monomer bis(2-hydroxyethyl terephthalate)(BHET) are both low, and the reaction rate is slow. Therefore, the development of new catalysts is still critical.

It has been found that synergetic effect of catalysts can speed up a chemical reaction or make reaction condition mild.[1,2] Yamamoto et al. found that combinations of organic and inorganic molecules bearing acidic groups could result in higher catalytic activity than individual catalysts for acid-catalyzed reactions.[2] Wang et al. and Zhou et al. found that the synergetic effect of cation and anion of ionic liquid catalyst could make the degradation condition of PET milder.[3,4] Therefore, we infer that a new catalyst containing two active components maybe has highly catalytic activity.

In this study, a metal-containing ionic liquid catalyst including two active components was synthesized and used for the PET degradation in ethylene glycol. The results showed that the synthesized catalyst has better low-temperature catalytic activity. The conversion of PET and the selectivity of BHET are 100 % and 78.7 % respectively at 175 Â°C for 1.25 h. The high catalytic activity is mainly attributed to the synergetic effect of two components. Moreover, this catalyst has the advantages of mild reaction condition, fast reaction rate, products being easily separated from catalyst and being of high purity.

[1] P. R. Schreiner. Science, 2010, 327, 965.

[2] H. Yamamoto and K. Futatsugi. Angew. Chem. Int. Ed., 2005, 44, 1924.

[3] H. Wang, R. Y. Yan, Z. X. Li, X. P. Zhang and S. J. Zhang. Catal. Commun., 2010, 11, 763.

[4] X. Y. Zhou, X. M. Lu, Q. Wang, M. L. Zhu and Z. X. Li. Pure Appl. Chem., 2012, 84, 789.

(478e) Anaerobic Digestion Technology:The Energy,Environment and Health of Developing Nations

Adetule, O., Cornell University

Originally presented on: 11/6/2013 13:50:00 - 14:10:00

Anaerobic digestion is one of the most cost effective and established technologies available for managing organic wastes for energy production,environmental management and pathogen reduction. In developing countries such as Nigeria, the application of this technology at household level as a method for gas generation and safe organic waste management has not been adequately studied. In light of the above, a pilot scale digester was designed for the production of Biogas which at the same time manages waste and reduced pathogens through anaerobic digestion process. The pilot scale digester was 10 litres and manages 9 litres of waste.The digester was fed with different types of waste:Cattle manure, Chicken litter, Rumen of cattle and Water hyacinth. After some weeks, the wastes produced Biogas, reduced pathogens and reduced the amount of wastes in the environment. A prototype digester was later designed and constructed which can accommodate more waste and produced more biogas. The prototype digester is a $5m^3$ underground fixed dome biodigester made of concrete and was designed, constructed and used for digesting cattle dung and water hyacinth. The digester was operated by mixing two parts of cattle dung (900kg) and one part of chopped water hyacinth (100kg) in 2,000 liters of water for two weeks following standard procedures. The gas produced was collected and piped to a kitchen for household use. It was also observed that there was reduction in pathogens and sanitization of the environment through waste management. This study shows that anaerobic digestion can be maximized for waste reduction, energy production,pathogen reduction and conversion of green house gases into usable forms.

(478f) Phosphorus Recovery From Municipal Wastewater Treatment Sludge

Lamichhane Upadhyaya, K., Mississippi State University
French, W. T., Mississippi State University
Hernandez, R., University of Louisiana at Lafayette
Originally presented on: 11/6/2013 14:10:00 - 14:30:00

Phosphorus recovery from municipal wastewater treatment sludge

Kamal Lamichhane Upadhyaya[1], W. Todd French[1] and Rafael Hernandez[2]

[1]Dave C. Swalm School of Chemical Engineering, Mississippi State University, MS, 39762, USA

[2]Department of Chemical Engineering, University of Louisiana at Lafayette, Lafayette, LA, 70503, USA

Abstract

Phosphorus is an essential nutrient for plants, animals and humans. Phosphorus is primarily used in agriculture, as it is one of the important components in fertilizer. Besides, phosphorus is also used in laundry detergent, matchstick, steel production and pesticides. The United States phosphate reserves are declining, which could lead to phosphorus shortage in near future. The United States now imports 10% of its phosphorus supply after exporting for several decades. Hence, there is need to search for renewable and sustainable source of phosphorus such as municipal wastewater sludge. Municipal wastewater sludge contains up to 10 wt% phosphorus and recovering these could significantly decrease the phosphorus shortage in the United States. In this study, the potential of phosphorus recovery from primary and secondary sludge through acidic and basic leaching processes will be investigated. The wastewater treatment sludge will be incinerated to obtain sewage sludge ash (SSA) and subjected to acidic and basic leaching at different acid and base concentration and pH levels. After leaching process, the supernatant will be separated from residue through filtration. Phosphorus will then be recovered from the supernatant mixture through the addition of calcium chloride in basic medium. The recovered phosphorus ($Ca_5(PO_4)_3OH$) has various agricultural and industrial usage as it resembles the phosphate mineral ore (apatite).

(478g) Techno- Economic Analysis of Wastewater Biosolids Gasification

Lumley, N. P. G., Colorado School of Mines
Porter, J. M., Colorado School of Mines
Originally presented on: 11/6/2013 14:30:00 - 14:50:00

Wastewater treatment in the U.S. produces over eight million dry tons of sludge per year. Wastewater sludge in its raw form is a noxious dilute suspension of organic matter in up to 95% water. Management and disposal is an expensive challenge for wastewater treatment plants (WWTPs), where the conventional approach to sludge management is to dewater the material to varying degrees and dispose by landfilling or application as a soil amendment. Such an approach requires energy and financial expense both in drying and ultimate disposal. Dried wastewater sludge has a lower heating value (LHV) close to that of low- rank coal. Thermochemical conversion of wastewater sludge to energy and/ or fuel product may allow for a more value- added approach to sludge management in addition to concentrating and stabilizing the waste stream.

Many WWTPs currently valorize a portion of the energy in sludge via anaerobic digestion. The biological process allows bacteria to operate on dilute sludge in an anaerobic environment producing a combustible bio- gas. Anaerobic digestion is an established technology but allows for only about 20% recovery of the energy in sludge. Anaerobic digestion does not significantly reduce the amount of sludge requiring downstream disposal. Additionally, bio -gas contaminants, especially siloxanes, are a problem for downstream combustion apparatus.

Thermochemical conversion (TCC) technologies subject sludge to chemical processes at elevated temperatures producing heat and/or combustible fuel products. TCCs applicable to wastewater treatment include direct combustion, pyrolysis, and gasification. TCC offers a great advantage over digestion by allowing nearly full utilization of sludge energy and a similarly great reduction in disposable waste volume. Each of these TCCs has been successfully applied to biomass. Wastewater sludge however is fundamentally different from most biomass sources. Raw wastewater sludge may be up to 95% water and 30% ash (dry basis). Balancing moisture removal with reactor operating conditions is the most important consideration in whether TCCs can be net energy- and net economically-positive. Ash concentration and composition present certain challenges to reactor design. Additionally, wastewater sludge organic composition can be variable leading to uncertainty in fuel output.

This work reviews the current state of practice of various TCCs with application to sewage sludge. We discuss technical and economic challenges restricting greater implementation of TCC within the wastewater process. A thermodynamic model of a sludge gasification- electric generator system is built using performance characteristics of a generic, commercially available small biomass gasification platform. We perform energy and techno- economic analyses on sludge gasification to determine lower- limit WWTP capacity and as- delivered sludge moisture content for both air- blown and steam- blown gasification processes.

(507a) Optimal Reactor Design for the Hydroformylation of Long Chain Olefins in Thermomorphic Solvent Systems

Hentschel, B.,
Peschel, A., Max Planck Institute for Dynamics of Complex Technical Systems
Freund, H., Max Planck Institute for Dynamics of Complex Technical Systems
Sundmacher, K., Max Planck Institute for Dynamics of Complex Technical Systems
Originally presented on: 11/6/2013 12:30:00 - 12:51:00

Optimal Reactor Design for the Hydroformylation of Long Chain Olefins in Thermomorphic Solvent Systems

Benjamin Hentschel[1], *Andreas Peschel*[2], *Hannsjörg Freund*[3], *Kai Sundmacher*[1,2]

[1]*Process Systems Engineering, University of Magdeburg, Magdeburg/DE*

[2]*Max Planck Institute for Dynamics of Complex Technical Systems, Magdeburg/DE*

[3]*Chemical Reaction Engineering, University of Erlangen-Nürnberg, Erlangen/DE*

The use of thermomorphic solvent (TMS) systems is a promising approach that is currently investigated regarding its applicability for the rhodium catalyzed hydroformylation of long chain olefins. By taking advantage of a temperature dependant miscibility gap, homogeneous reaction and efficient liquid-liquid phase (and thus catalyst) separation can be performed with a single solvent system by using the temperature as "switch" to tune the phase behavior.

On the example of the hydroformylation of 1-dodecene in a TMS it is demonstrated how the optimal reactor for such innovative systems can be derived. The demanding interplay of the thermodynamics of the TMS, complex reaction network, and solvent recycles requires a model based reactor design approach which considers the dynamics of the whole process rather than an isolated reactor design. For this purpose a reference process flow sheet (Fig. 1) was developed in order to allow for the derivation of a process-wide optimal reactor.

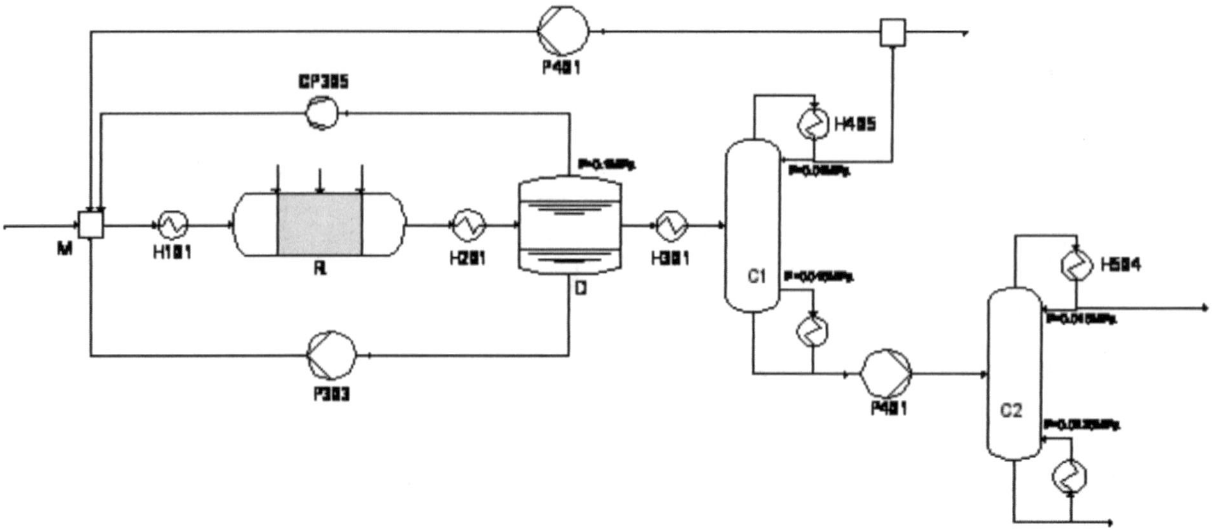

Fig. 1: Flow sheet of the hydroformylation process.

Due to the numerous process constraints and the complexity of the reaction system a heuristic reactor design approach is unlikely to be able to yield the optimal reactor. Hence, to account for the full potential of this innovative reaction and solvent system a reactor design based on the optimal reaction route is pursued. Considering a fluid element of the system which is traveling through the reactor, optimal profiles of energy and material fluxes over reaction time can be calculated as solution of an optimal control problem. This problem is constrained by the dynamic equations of the fluid element, constitutive equations describing phase properties, reaction kinetics, gas solubilities, and the balance equations of the remaining process units. For each unit appropriate cost models are implemented enabling the minimization of the total production costs of the desired linear aldehyde product.

As a result of this optimization it could be shown that the solvent recycle leads to an increase in the selectivity that can be achieved in the reactor. Thus, a separate optimization of the reactor without considering the recycles would lead to a falsified lower reactor performance prediction. Furthermore, the cost optimal solution, considering all recycles, turns out to be a combination of necessary separation efficiency and optimal reaction route. Assuming an annual production of 10.000 tons of product, the advanced reaction route leads to a reduction of the total production costs of 5% compared to an optimized reference case. This corresponds to an annual saving of about 2.2 Mio. US$.

In summary it could be demonstrated that reactor and process have to be optimized as an integrated system in order to find the process-wide optimal solution, rather than focusing on local objectives of the reactor, such as product selectivity. Exemplified on the hydroformylation of long chain olefins in TMS it could be shown that the presented reactor design method can yield a process-wide optimal reactor design based on the cost optimal solution, in which optimal reactant dosing and heat removal is realized along the reactor.

(507b) A Framework for Process Synthesis Integrated With Sustainability and Process Intensification

Babi, D. K., Denmark Technical University
Holtbruegge, J., Department of Biochemical and Chemical Engineering, Laboratory of Fluid Separations
Lutze, P., TU Dortmund University
Woodley, J., Technical University of Denmark
Gorak, A., Chair of Fluid Separation Processes
Gani, R., Technical University of Denmark
Originally presented on: 11/6/2013 12:51:00 - 13:12:00

A Framework for Process Synthesis integrated with Sustainability and Process Intensification

Deenesh K. Babi[1], Johannes Holtbruegge[2], Philip Lutze[2], John M. Woodley[1], Andrzej GÃ³rak[2], Rafiqul Gani[1]

dkbabi@kt.dtu.dk, johannes.holtbruegge@bci.tu-dortmund.de, philip.lutze@bci.tu-dortmund.de, jw@kt.dtu.dk, andrzej.gorak@bci.tu-dortmund.de, rag@kt.dtu.dk

[1]*Department of Chemical and Biochemical Engineering, Technical University of Denmark (DTU), DK-2800 Kongens Lyngby, Denmark*

[2]*Department of Chemical and Biochemical Engineering, Technical University of Dortmund (TU Dortmund), D-44227 Emil-Figge-Str. 70 Dortmund*

The objective of process synthesis is to find the best flowsheet, among numerous alternatives for converting specific raw materials to specific desired products subject to predefined performance criteria. The unit operations concept is the one most used for process synthesis because it allows to associate the operational tasks with the processing route to be followed. This concept has been sufficient until now, however, due to increasing demands on the chemical industry, such as the design of sustainable/new processes, an extension of the current concepts by considering lower levels of aggregation than the unit operations level as well as sustainability is necessary. That is, when applying the unit operations concept one automatically connects the unit operations with their function and therefore this mode of thinking restricts the solution space to the limited number of ?'well defined" unit operations [1]. However, by incorporating a sustainability analysis within process synthesis bottlenecks can be identified and ordered in terms of those having the highest impact for improvement of the process with respect to the environment. But within the solution space, intensified and integrated unit operations must also be included in order to achieve a targeted improvement. Therefore, by incorporating different levels of aggregation at the flowsheet, the unit operations, the task and the phenomena levels, the opportunity to innovate is provided.

Different methods have been developed which perform process synthesis incorporating process intensification at the different levels [2, 3]. This work focuses on the inclusion of a sustainability analysis as a means by which process bottlenecks can be identified and the inclusion of the phenomena level [4] as a lower level of aggregation in process synthesis. It should be noted that the degree of complexity as well as the search space increases as one operates at the lower levels of aggregation. For example, the considered process phenomena need to be combined to perform the identified tasks and these combinations are then translated into unit operations, which define the final process flowsheet. Therefore, operating at the phenomena level allows one to include known as well as new intensified unit operations as part of process synthesis.

The computer-aided framework for process synthesis has been developed and it consists of 10 hierarchical steps exclusive of the intensification method. Steps 1-6 consist of the problem and objective function definition, the investigation of whether a base-case design exist or not. If no, flowsheet alternatives are generated and screened using the means-ends analysis [2] and thermodynamic insights [5]. If yes, then this design is considered as a reference for further improvement/intensification. Steps 7-8 consist of performing a sustainability analysis using data obtained from rigorous simulation of the base case design, to identify process bottlenecks. Having this information the phenomena-based synthesis (PBS) intensification sub-method is entered. Steps 9-10 consist of equipment sizing and economic evaluation followed by a further environmental and sustainability analysis for comparison to the base case design.

The PBS sub-method consists of 7 hierarchical steps. Steps 1-2 consists of metric definition and translation which provide the system constraints related to structural, logical and operational and process analysis where the system is decomposed from the flowsheet level to the phenomena level. The thermodynamic properties of the system [4] are also investigated in order to identify further process bottlenecks. Steps 3-4 consists of identifying tasks and phenomena for overcoming the process bottlenecks based on the thermodynamic insights gained from the process analysis. The means-ends analysis together with the identified phenomena which are connected, are used for generating the flowsheet alternatives based on tasks and phenomena. These connected phenomena referred to as operations are then translated to unit operations which fulfil process requirements, whether traditional, existing or innovative PI equipment. In translating the connected phenomena to unit operations a concept known as S-Files, similar to SMILES for representing molecular information, is used in order to store and easily retrieve from a knowledge-base the unit operations that built the flowsheet. In case a unit operation cannot be found from a set of connected phenomena then a design is proposed for a new unit operation. Steps 5-7 consist of stepwise screening the alternatives with respect to the constraints set previously in Step 2 and optimizing for identifying the most promising flowsheet.

In the proposed presentation the computer-aided framework will be presented along with the tools and data needed for its application, which will be highlighted through two case studies: the production of methyl-acetate for verification of the framework and the production of di-methyl-carbonate as an industrial example.

References

[1] H. Freund, K. Sundmacher. Ullmann's Encyclopedia of Industrial Chemistry, Process Intensification, 2. Fundamentals and Molecular Level. Wiley-VCH Verlag GmbH & Co. KGaA, 2000

[2] Jeffrey J. Siirola. Strategic process synthesis: Advances in the hierarchical approach. Comput. Chem. Eng. 1996, Supplement 2 (20) S1637-S1643

[3] Lutze, A. RomÃ¡n-Martinez, J. M. Woodley & R. Gani. A systematic synthesis and design methodology to achieve process intensification in (bio) chemical processes. Comput. Chem. Eng. 2012 (36) 189? 207

[4] P. Lutze, D. K. Babi, J. M. Woodley, and R. Gani. Phenomena Based Methodology for Process Synthesis Incorporating Process Intensification. Ind. Eng. Chem. Res. 2013 Special Issue PSE-2012

[5] C. Jaksland, R. Gani, K. M. Lien. Separation process design and synthesis based on thermodynamic insights. Chem Eng Sci, 1995 (50) 511-530

(507c) Theoretical Investigation of a Semicontinuous Distillation Process for the Separation of Bio-Based Impurities

Niesbach, A., Laboratory of Fluid Separations
Adams, T. A. II, McMaster University
Lutze, P., TU Dortmund University
Originally presented on: 11/6/2013 13:12:00 - 13:33:00

The transition from fossil-based raw materials to bio-based raw materials is necessary in order to produce chemical products in a sustainable way. For example, *n*-butyl acrylate is currently produced through a synthetic route using chemicals derived from petroleum, but new research into its production from biological sources (such as wood) is showing promise as a more sustainable alternative. However, the composition of the impurities contained in bio-derived *n*-butyl-acrylate (and its precursors) significantly differs from traditional, fossil-derived *n*-butyl acrylate. Furthermore both the composition and the amount of impurities in bio-derived *n*-butyl acrylate change depending on the biomass feedstock used and even from natural batch-to-batch variations in the biosynthetic reactions. Therefore highly flexible processes which can achieve the required final product purity even with varying initial impurity concentrations need to be developed.

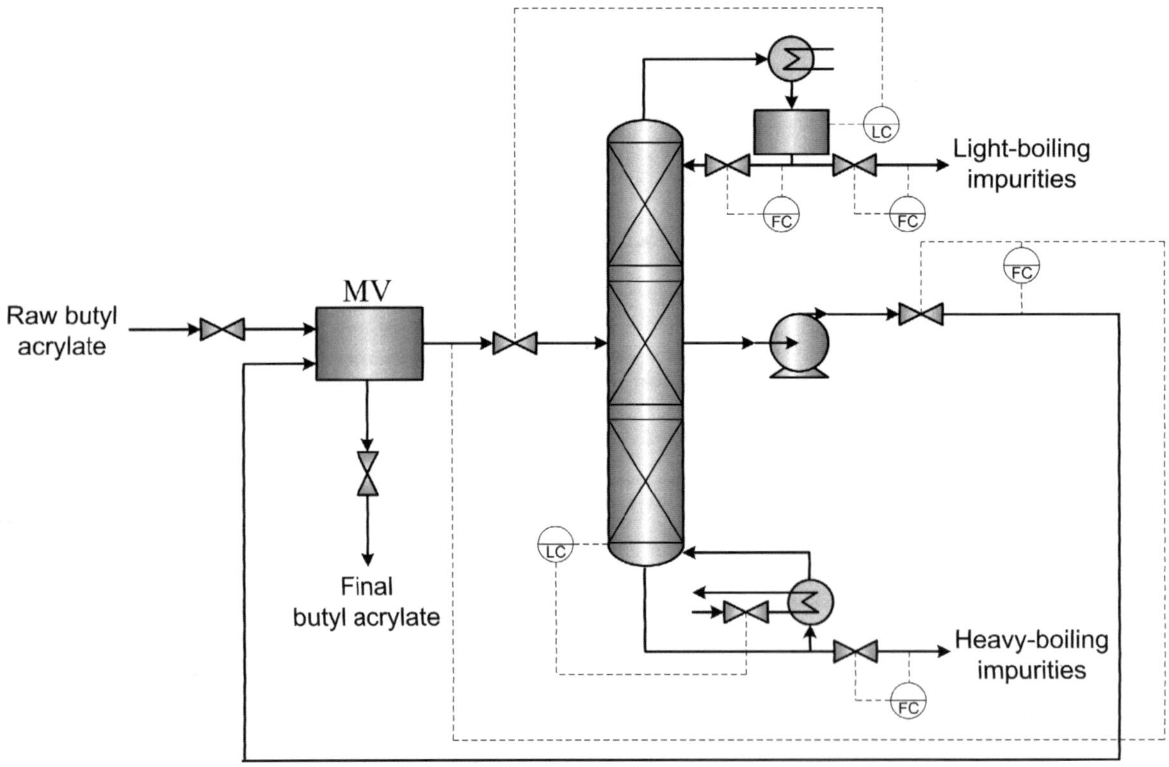

Figure 1: Configuration of semicontinuous distillation process.

In a previous study, a reactive distillation column for the esterification reaction of bio-*n*-butanol and bio-acrylic acid to produce *n*-butyl acrylate was investigated, which is a promising candidate to replace the conventional process, reducing the production costs for *n*-butyl acrylate. Besides the development of the reactive distillation process, the potential impurities resulting from the biosynthetic production of the reactants were identified. However, the use of bio-based raw materials resulted in insufficient final product purities. Therefore, new and flexible downstream process concepts are investigated in this work in order to remove the impurities.

A promising process concept for the purification of bio-based chemicals is semicontinuous distillation. Semicontinuous distillation processes allow for the separation of three or more streams simultaneously by tightly integrating one or more middle vessels to a distillation column. In this study, the application of a ternary semicontinuous distillation process for the removal of bio-based impurities is studied for the purification of three different material streams: the purification of high-purity bio-based *n*-butyl acrylate (see Fig. 1) as well as for the purification of bio-*n*-butanol and bio-acrylic acid which are the reactants fed to the reactive distillation column which produces *n*-butyl acrylate. The simulation studies are performed using Aspen Dynamics? V7.3. Three performance criteria which are the specific heat study, the specific product loss and the cycle time of the semicontinuous process are used to characterize and compare the feasibility of the different case-studies. Simulations with varying initial and final impurity concentrations are performed to demonstrate a high flexibility. It is shown that the semicontinuous technology is feasible for two of the investigated case-studies, the purification of bio-*n*-butanol and the purification of the final product *n*-butyl acrylate. Purities of up to 99.95 wt.-% are achieved with varying initial purities of 99.00 wt.-% or higher. Different impurity compositions can be handled by this technology and varying final purities are achieved by only changing the cycle time of the process.

Hence, this technology shows a high potential for the purification of bio-based chemicals, offering a high flexibility in respect to initial and final impurity concentrations. Furthermore, it should be noted that the same semicontinuous set-up (meaning the same column height and diameter) used to remove bio-based impurities from the final product *n*-butyl acrylate can also be used to remove impurities from the reactants bio-*n*-butanol and bio-acrylic acid. This demonstrates the potential of semicontinuous systems to be used as a multipurpose column which may reduce capital costs.

Acknowledgement: The research leading to these results has received funding from the European Union Seventh Framework Programme (FP7/2007-2013) under grant agreement nÂ° 241718, EuroBioRef. Financial support of the Martin-SchmeiÃŸer-Stiftung is gratefully acknowledged.

(507d) Integration of Controllability and Diagnosticability of Reactive Distillation Column As From Conceptual Design Step. Application to the Production of Ethyl Acetate

Fernandez, M. F., Polytechnic School of the University of Sao Paulo
Meyer, X. M.,
Le Lann, M. V., CNRS
Huynh, N.,
Brehelin, M.,
Le Roux, G. A. C., Polytechnic School of the University of Sao Paulo
Originally presented on: 11/6/2013 13:33:00 - 13:54:00

Increasing attention has been directed towards reactive distillation processes as a successful process integration example. Reactive distillation means the simultaneous implementation of reaction and distillation in a counter currently operated column, where chemical equilibrium is superimposed on vapor liquid equilibrium. Conversion can be increased far beyond what is expected by the equilibrium due to the continuous removal of reaction products from the reactive zone, reducing capital and investment costs and contributing for sustainable development due to a lower consumption of resources. Reactive distillation process is although a complex system; the combination of separation and reaction zones leads to strong interactions and nonlinearities, hindering the process control and diagnosis. When the conceptual design is optimally obtained based on a static analysis, regarding only economics and environmental criteria, the result may be a unit very difficult to control. It is thus appropriate to consider controllability and diagnosticability criteria as from conceptual design steps.

An entire methodology for the conceptual design of a reactive distillation column that considers controllability and diagnosticability aspects is proposed and applied to the production of ethyl acetate from esterification of acetic acid and ethanol.

A tool of feasibility and pre-design of reactive columns has been developed in the LGC (Laboratoire de GÃ©nie Chimique, Toulouse, France) based on a sequential approach with development and application of different types of software (Thery et al. 2005). First, a combination of algorithms developed in Fortran and MatLab® connected to a thermodynamic properties server (Simulis®) is used for feasibility and synthesis calculations. Then, the column configuration is defined in Aspen Plus®and process simulation at steady state provides more operational parameters, by considering the heat phenomena. The main interest is that these tools lie on sequential and progressive introduction of process complexity; from a minimal set of information concerning the physicochemical properties, successive refinements are considered. The techniques applied on the feasibility analysis, the synthesis and the conceptual design steps lead to the project of a column configuration and the necessary operating parameters to attain the process objectives. Different configurations may be feasible; so that the engineer should choose the best configuration according to his own aims (lower investment, lower operating costs, better operability, etc.).

The objective now is to choose the column with best operability and to identify the structural parameters that provide the best controllability. By considering an indices-based analysis method, some criteria to quantify and qualify the system controllability are identified at early design steps. The first steps of the controllability methodology are obtained by steady state simulations, with the calculation of the Sensitivity Matrix, the Singular Value Decomposition, the Condition Number and the Intersivity Index (Moore 1992). Then, the dynamic simulation is studied, by considering the properly chosen control loops with the Integral of the Squared Errors calculations. At this moment, the importance of an experimental validation is highlighted so as to provide realistic hydrodynamic parameters, to understand the sensitive parameters such as heat losses and to adapt values for the catalyst behavior in function of the system.

Conventional distillation columns have two degrees of freedom, so two different variables have to be controlled to ensure good separation and the respect of production quality. In the case of reactive distillation columns, a desired conversion rate is also of control interest. The additional degree of freedom given by the feed ratio on a double-feed column configuration allows the introduction of a third control loop with a measured variable inside the reaction zone that addresses well the reaction conditions. The three control loops are defined so as to show high sensitivity, good balances and small interferences. The application of the methodology for the production of ethyl acetate identifies the system specific sensibilities and the best control configurations. It is found that the addition of theoretical stages between the two feed positions ameliorates the column controllability and that homogeneous catalysis results on more reliable operability than heterogeneous catalysis.

Once the column configuration is proposed and the process dynamic simulation is mastered, the objective of process diagnosis is to rapidly detect abnormal operating conditions so as to avoid reduction on its performance and to promote proper operation with more security to its surroundings. The classification technique LAMDA (Learning Algorithm for Multivariable Data Analysis), based on fuzzy logic and developed at the LAAS (Laboratoire d'Analyse et d'Architecture des Systèmes, Toulouse, France), is considered and the calculations are made with the software P3S®(Kempowsky et al. 2003). The simulation data with regard to potential operating faults lead to a learning procedure, when the knowledge of human experts is included; the algorithm is applied and the historic measures are differentiated into several classes, which are to be interpreted as physical operating situations by the human expert. After the learning procedure, a recognition step is lead: the data is analyzed, the classes already established can be identified and the most relevant sensors are identified automatically by the Membas procedure based on their ability of optimizing the fault discrimination (Molina 2005). The composition sensors were verified to give strong information of the process diagnosis. However, the complexity on using online composition analyzers can be overcome by the use of a higher quantity of temperature sensors to give the same situation recognition degree. The diagnosticability analysis at the design phase of processes is important so as to preview the installation of relevant sensors at specific column locations. Some sensor placements are not always possible when the column has a fixed structure and the lack of measures would jeopardize the system operation.

In conclusion, some indicators of controllability and diagnosticability have been identified to be considered as from the design step of reactive distillation columns. The work results on important contributions to the study of the ethyl acetate production by reactive distillation, due to the fact that few studies address the dynamic behavior of this system.

References:

Kempowsky T., Aguilar-Martin J., Subias A. et Le Lann M.V. Classification Tool based on Interactivity between Expertise and Self-Learning Techniques, IFAC-Safeprocess, Washington D.C., USA, 2003.

Molina A. O. MÃ©thodologie pour le placement des capteurs a base de mÃ©thodes de classification en vue du diagnostic. PhD thesis, Laboratoire d'Analyse et d'Architecture des Systèmes du CNRS, France, 2005.

Moore C. Chapter 8, Practical Distillation Control. Van Nostrand Reinhold, edited by William L. Luyben, 1992.

ThÃ©ry R., Meyer X. M., Joulia X. et Meyer M. Preliminary design of reactive distillation columns. *Chemical Engineering Research and Design*, 83 (A4), 379, 2005.

(507e) Optimization of a Reactive Distillation Process for Producing Triethyl Citrate

Orjuela, A., Universidad Nacional de Colombia
Martinez, C. A., Universidad Nacional de Colombia
Molina, E. J.,
Jimenez Cucaita, L. E.,
Originally presented on: 11/6/2013 13:54:00 - 14:15:00

Triethyl citrate is a bio-based plasticizer mostly used in polymers that may have contact with humans (medical devices, toys, etc). Currently it is produced by esterification of citric acid with ethanol in a batch reactor system followed by a striping process to remove excess of ethanol. In order to reduce energy and materials consumption of the traditional process, reactive distillation (RD) can be implemented. In this process high yield triethyl citrate can be obtained as bottom product while ethanol excess and water are removed as distillate. Some authors reported the feasibility of this process, however low conversion to triethyl citrate and high energy requirements were observed.

In this work it was developed and optimized a RD process for triethyl citrate. By reprocessing reported kinetic data a pseudo homogeneous model was obtained. NRTL-HOC activity based equation was used for phase equilibria modeling. The method used for the optimization corresponds to a stochastic differential evolution, setting the objective function to minimize energy consumption and the purity of product streams. This optimization was automated by connecting Aspen One^{TM} and Visual Basic^{TM} included in EXCEL^{TM}. A reduction in energy consumption of around 30% with respect to conventional reaction followed by distillation process was observed.

(507f) Reactive Separation of Formic Acid, Acetic Acid and Water From Biorefinery Feed

Painer, D., Graz University of Technology
Lux, S., Graz University of Technology
Siebenhofer, M., Graz University of Technology
Originally presented on: 11/6/2013 14:15:00 - 14:36:00

In the biorefinery valuable bulk products are rarely provided in simple mixtures of appropriate composition. Formic acid/acetic acid/water mixtures from pulp industry provide an impressive demonstration for non-separable systems from the distillation point of view. Isolation of highly pure acids is challenging since conventional rectification is more or less impossible, because of facing a binary high boiling azeotrope and a ternary saddle azeotrope.

Reactive separations give access to a variety of promising process concepts which allow effective product isolation from complex mixtures with thermodynamic separation limits. Separate esterification of formic acid and acetic acid are well investigated chemical reactions. However, the combination of esterification with distillation of either both carboxylic acids or selective esterification with distillation give access to a new approach for reactive separation of azeotropic mixtures.

Esterification with methanol will form low boiling esters ready for complete separation by distillation. Permanent removal of the esters shifts the equilibrium towards complete conversion of acids. Because of intramolecular catalysis no further catalyst is needed.

In a first step the vapor/liquid-equilibrium of the ternary system was verified experimentally at ambient pressure. For simulation of vapor/liquid equilibrium data the UNIQUAC model and the Hayden O'Connell equation were applied.

During process development the reaction kinetics of the individual esterification reactions were then investigated and kinetics was modeled. Simultaneous esterification of acetic acid and formic acid was then validated experimentally. Because of faster esterification of formic acid, the amount of methyl formate outweighs methyl acetate formation. Along with the difference in the boiling points of the formate and the acetate this gives access to selective removal of formic acid from the mixture.

Product isolation routines were validated with batch rectification, confirming highly selective product separation. Validation was completed with different feed compositions to demonstrate that the distillation limits of the ternary system may be avoided by chemically altering the substance properties, enabling a flexible technological toolbox for process design.

(507g) Design of Reactive Separation Process for CO2 Sequestration From Exhaust Gases of Process Plants

Iyer, S. S.,
Mantravadi, V. K., Indian Institute of Technology Madras
Renganathan, T., Anna University
Kaisare, N., Indian Institute of Technology Madras
Pushpavanam, S., Indian Institute of Technology Madras
Originally presented on: 11/6/2013 14:36:00 - 14:57:00

Methanol production by catalytic reduction of carbon dioxide is an option for carbon sequestration to reduce the carbon footprint of power plants and other process industries. Pre- and post-combustion capture of carbon dioxide has been a big area of research. For example, in coal fired thermal power plants, gasification of coal is used for pre-combustion capture carbon dioxide. Chemical absorption processes using mixtures of amines are well established for post-combustion capture. Of various technologies for CO_2 sequestration, reduction with hydrogen to produce methanol is of interest to us.

Hydrogen is mostly produced from the natural gas. Electrolysis of water, though an energy intensive alternative, is also used for production of hydrogen. Hydrogen so obtained can be used for catalytic hydrogenation for CO_2 sequestration. The catalytic hydrogenation of CO_2 over promoted Cu-ZnO catalysts under pressurized conditions produces mainly CH_3OH, CO and H_2O [1, 2]. At higher temperature, a small amount of CH_3OCH_3 (as well as some methane) may also be produced. Therefore, main reactions are shown in the following equations: (a) $CO_2 + 3H_2 \leftrightarrow CH_3OH + H_2O$ and (b) $CO_2 + H_2 \leftrightarrow CO + H_2O$. Increase in selectivity could also be due to decrease in CO yield through the reaction of CO with hydrogen to produce methanol (c) $CO + 2H_2 \leftrightarrow CH_3OH$.

This work will highlight a range of different reactive separation processes for catalytic synthesis of methanol for sequestering carbon dioxide. The key objective is to design and optimize a viable option for catalytic conversion of CO_2 to methanol on Cu/ZnO based catalyst. The traditional design or optimization method is to increase the productivity of methanol. However, the objective here is to meeting the environmental regulation of CO_2 emission by sequestration. Different approaches by which we can enhance the conversion of CO_2 will be studied. Some of them include, (1) manipulating the temperature / pressure profiles in order to drive the reaction for methanol synthesis depriving other reactions [1], and (2) removal of methanol as soon as it is formed while maintaining the conditions favorable for methanol synthesis.

Keywords: CO_2 Sequestration, Thermodynamic analysis, Reactive separation, Kinetic analysis

References

[1] H. Arakawa, J.-L. Dubois, and K. Sayama, Selective conversion of CO_2 to methanol by catalytic hydrogenation over promoted copper catalyst, *Energy Convers. Mgmt*, 33 (5-8), 1992, 521-528.

[2] K. Takeshi, H. Itaru, M. Hirotaka, M. Kozo, U. Kenji, W. Taiki, and S. Masahiro, Kinetic study of methanol synthesis from carbon dioxide and hydrogen, *Applied Organometal. Chem.*, 15, 2001, 121-126

(534a) Plate Food Waste: A Valuable Source of Energy

Jung, S., Iowa State University

Originally presented on: 11/6/2013 15:15:00 - 15:35:00

Post-consumer food waste, because of its quantity and composition, could be a good feedstock for biofuel production. The effect of glucan loading (2 and 6%), addition of an enzyme cocktail, pretreatment (grinding/drying or sterilizing), and food waste composition (breakfast vs. lunch, presence of used napkins) on ethanol concentration was reported. The potential for sorting and converting used napkins into ethanol was also assessed. At a 2% glucan content, ethanol concentration of up to 16 g/l was obtained after saccharification/fermentation of breakfast and lunch food waste. Increasing glucan content to 6% significantly increased ethanol concentration to 54 g/L, while there was no significant effect of food waste composition on carbohydrate conversion efficiency. When converting ethanol concentration from g/L to L/g of waste, concentrations of 0.06, 0.06 and 0.40 were obtained for food waste, food waste w/o napkins, and used napkins, respectively, identifying used napkins as an excellent source of bioethanol production.

(534b) Biofuel Production and Kinetics Study of Catalytic Microwave Pyrolysis of Douglas Fir Pellet Over Activated Carbon Supported Metal Catalyst

Bu, Q., Washington State University
Lei, H., Washington State University
Wang, L., Washington State University
Tang, J.,
Originally presented on: 11/6/2013 15:35:00 - 15:55:00

The presented study is aiming to improve the quality of bio-oils by catalytic upgrading of pyrolysis oil from microwave pyrolysis of biomass using activated carbon (AC) supported metal catalyst. A central composite experimental design (CCD) was used to optimize the reaction condition of this process. The effects of reaction temperature and reaction time on product yields were investigated. GC/MS analysis indicated that the main chemical compounds of bio-oils were furans, phenols, guaiacols, and ketones/ethers which were in the range of 80-87% of bio-oils for most of the samples. The ketones/ethers mainly composed by 2-cyclopen-1-one and straight chain ethers were increased up to about 38% of bio-oil, and the phenols increased significantly accompanying with the decrease of guaiacols in comparison with raw biomass pyrolysis. Thermogrametric analysis (TGA) was used to analyze the thermal behavior or Douglas fir pyrolysis, and the kinetic parameters were investigated using thermogrametric analysis (TGA) data and compared with catalytic microwave pyrolysis of Douglas fir in this process, the activation energy of Douglas fir pyrolysis in this process was remarkable lower than conventional TGA pyrolysis of Douglas fir. And the reaction mechanism of this process was analyzed.

(534c) Molecular-Level Modeling of Municipal Solid Waste Gasification

Horton, S. R., University of Delaware
Zhang, Y., Air Products and Chemicals, Inc.
Bennett, C. A., University of Delaware
Klein, M. T., University of Delaware
Petrocelli, F., Air Products and Chemicals, Inc
Originally presented on: 11/6/2013 15:55:00 - 16:15:00

Molecular-level Modeling of Municipal Solid Waste Gasification

Scott R. Horton*, Craig A. Bennett, Michael T. Klein

Department of Chemical & Biomolecular Engineering, University of Delaware

Yu Zhang, Francis P. Petrocelli

Air Products and Chemicals, Inc.

In 2010, there was 250 million tons of municipal solid waste (MSW) in the United States. Over half of this waste ended up in landfills which fail to recover the energy stored within MSW. The most basic waste-to-energy technology is incineration. While this process recovers energy, it also poses problems to the environment due to SO_x, NO_x, dioxin, and furan emissions. Plasma-arc gasification has nonstoichiometric amounts of oxygen and reduces the production of these harmful oxygenates while recovering energy from waste via syngas production. Furthermore, due to the extremely high temperatures of the plasma torch, the final by-product is a salable glass-like slag. The objective of this study is to aid process design by developing a molecular-level kinetics model of plasma-arc gasification.

The MSW is typically described at the lumped level, i.e., biomass, food waste, and plastics. In the present work, each of these lumped components was described at the molecular-level. Biomass was modeled as cellulose, hemicellulose, and lignin. In addition to these molecules, lipids, common amino acids, and sugars were used to model food waste. The modeled plastics were polyethylene, polyethylene terephthalate, and polyvinyl chloride. These feeds were all represented at the molecular level using Flory statistics for nearly linear polymers and lattice theory for lignin.

The feed stream was converted into a reaction network using an in-house software tool, the Interactive Network Generator (INGen). There were two major categories of reactions: devolatilization/pyrolysis and gasification. The major pyrolysis reaction families were thermal cracking, decarboxylation, and decarbonylation. For gasification, the dominant reaction type was reaction with oxygen. The generated reaction network was solved and optimized using another in-house software, the Kinetic Modeling Editor (KME).

For integration with process simulation, the optimized KME model was integrated into ASPEN PLUS using user-defined models. In the case of gasification, the KME model contains reactors in series and rate laws employing linear-free energy relationships for reaction families. The process of KME-Aspen integration was generalized allowing for the automatic implementation of any reaction model into an ASPEN worksheet.

The KME model could also be coupled to a computational fluid dynamics (CFD) solver and take equipment level information into account in the design. Once the products and kinetics of the pyrolysis and gasification reactions are obtained from KME, the reaction network could be converted into one or several global pyrolysis/gasification reactions by kinetic model reduction. The simplified mechanism could be easily integrated into any CFD solver, such as Ansys FLUENT. On the other hand, equipment information obtained from the CFD simulation, such as flow and temperature pattern, provides guidance for choosing reactors in ASPEN PLUS process simulation.

(534d) Performance Analysis of Rdf Gasification in a Two Stage Fluid Bed - Plasma Process

Materazzi, M., University College London

Lettieri, P., University College London

Chapman, C., Advanced Plasma Power

Taylor, R.,

Originally presented on: 11/6/2013 16:15:00 - 16:35:00

PERFORMANCE ANALYSIS OF RDF GASIFICATION IN A TWO STAGE FLUID BED - PLASMA PROCESS

M. MATERAZZI* **, P. LETTIERI*, R. TAYLOR **, C. CHAPMAN **

** Department of Chemical Engineering, University College London, Torrington Place, London WC1E 7JE, UK*

*** Advanced Plasma Power, South Marston Business Park, Swindon, SN3 4DE, UK*

Tar generation and ash disposal represent the strongest barrier for use of single stage fluid bed gasification for waste treatment, whereas sufficing for both is only possible with expensive cleaning systems and further processing. Waste fuels have, in general, physical and chemical characteristics much different from fossil fuels, so that the behaviour of these two feedstock types in high temperature fluid beds is very different one from the other. The high content of ashes and inorganic materials in RDF (Refuse Derived Fuel) dictates the thermal conditions in the fluid bed. A relatively low temperature has to be used to prevent agglomeration and sintering of bed material. For this reason, the gas that is produced by a standard fluid bed gasifier (FBG) has tars and other condensable organic species that are technically difficult and costly to remove. Furthermore, the bottom ash that is generated in the reactor may contain high levels of carbon, heavy metals and organic pollutants which lower the conversion efficiency of the process and limit any secondary usage.

The use of plasma systems has increasingly been applied with thermal waste treatment for its ability to completely decompose the input waste material into a tar-free synthetic gas and an inert, environmentally stable, vitreous material known as slag. The principal advantages that plasma offers to thermal conversion processes, besides the already mentioned tar/ash related issues absence, are a smaller installation size for a given waste throughput, and the use of electricity as energy source, characteristics which permit the technology to treat a wide range of low calorific value materials including liquids and solids. In applying the plasma technology to the gaseous products from a FBG, an advanced two-stage thermal process is able to achieve efficient conversion of the chars and the complex organics to the primary syngas constituents whilst limiting the electrical energy demand of the process (Figure 1).

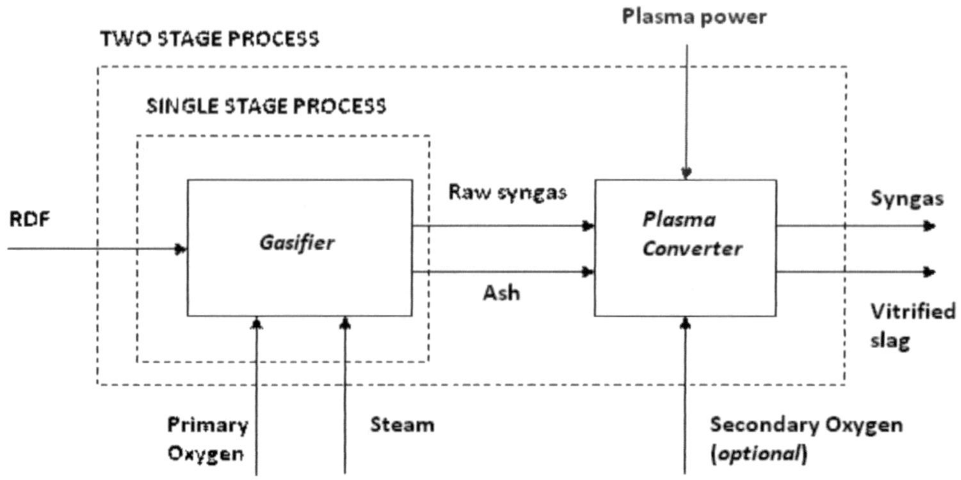

Figure 1: Schematic of Gasplasma two-stage thermal process.

This study focused on the thermodynamic assets of using a two-stage (FBG + plasma) thermal process over the conventional single-stage approach. These include, for example, the fact that the primary thermal waste decomposition is performed in conditions of optimal stoichiometric ratio for the gasification reactants. Furthermore, staging the oxidant injection in two separate intakes significantly improves the efficiency of the system, reducing the plasma power consumption. The aim of the present work is also to determine the effects of the thermal parameters so as to be able to select optimum gasification conditions in a two stage process and to clarify the mechanism of thermal decomposition of carbonaceous materials in the FBG and in the plasma converter.

A flexible model capable of providing reliable quantitative predictions of product yield and composition after the two-stage process has been developed. The method has a systematic structure that embraces atom conservation principles and equilibrium calculation routines, considering all the conversion stages that lead from the initial waste feed to final products. The model was also validated with experimental data from a demonstration plant in Swindon (UK) (Table 1).

	Case 1	**Case 2**	**Case 3**	**Case 4**	**Case 5**
Description:					
O2/fuel ratio (w/w)	0.51	0.59	0.59	0.79	0.50

Bed temperature (°C)	770	720	795	720	800
Proximate analysis, % (w/w)					
Fixed carbon	6.4	12.2	11.6	8.5	22.8
Volatile matter	59.6	50.2	64.8	47.6	68.0
Ash	19.1	23.2	12.1	8.9	0.5
Moisture	14.9	14.4	11.5	35.0	8.7
Ultimate analysis, % (w/w)					
C	41.0	47.0	43.0	31.5	45.2
H	5.7	6.3	5.6	4.1	6.46
O	17.5	6.9	26.6	19.7	45.38
N	1.2	1.74	0.61	0.4	0.26
S	0.2	0.15	0.25	0.17	0.01
Cl	0.4	0.31	0.34	0.23	0.25
GCV, MJ/kg (dry basis)	22.1	26.4	21.0	21.0	22.0

Table 1: Experimental parameters and characteristics of solid wastes (as received).

The study effectively demonstrated that the two-stage gasification system significantly improves the gas yield of the system and the carbon conversion efficiency, which are crucial in other single stage FBG systems, whilst maintaining high energy performances.

The most important outcome of this analysis is that the important parameters that define the performance of the fluid bed-plasma process are the oxidants (oxygen and/or steam) streams and the plasma power. These parameters are function of the type of waste and the detailed aim of the project. Furthermore, the two-stage performance can be quite accurately predicted by assumption of thermodynamic equilibrium.

Figure 2: Comparison of model results with experiment data from the (left) FBG gasifier; (right) FBG + Plasma converter

There is close correlation between the observed and predicted values of the syngas exiting the plasma converter, although the actual level of CO is still ~1% below the theoretical value while the CO_2 level is 1.5% higher. The noted differences are likely due to flow and RDF composition uncertainties rather than from the predictive model. Given the size and configuration of the demonstration plant and the inherent variability of real RDF uncertainties are to be expected. The layout of syngas heating value (LHV) and cold gas efficiency (CGE) have been calculated and analyzed. It was found out that high LHV and CGE values are maintained in different power and oxygen conditions. The reason is that addition of plasma power into the converter decreases the amount of secondary oxygen required for complete gasification and produces larger amounts of CO and H_2 in the product gas. The optimizing direction for the two-stage process can only be determined after considering the detailed situation on different projects.

(534e) Process Modeling of Continuous Catalytic Gasification As a Waste to Energy Alternative

Tillie, C. F., Cleveland State University
Riedthaler, M. J., Cleveland State University
Reeves, S. A., Cleveland State University
Adhikari, S., Cleveland State University
Gatica, J. E., Cleveland State University
Originally presented on: 11/6/2013 16:35:00 - 16:55:00

Presently, space exploration waste is either carried onboard for the duration of the mission or discarded into a trash module which is sent back into Earth's atmosphere for destruction. On the other hand, with an estimated cost of $10,000 per pound, the cost of delivering mass to locations beyond low earth orbit (LEO) locations is prohibitive.

Since approximately 15 pounds of waste are produced daily by a crew of four during a mission, the intrinsic value of this waste can be greatly enhanced by waste to energy technologies. Moreover, conceptualizing waste gasification as an in-situ alternative for processing of trash is also an efficient route to control waste, while maintaining a healthy habitable environment, during long-duration missions (Santiago-Maldonado et al., 2010).

The gasification of long-chain polymers is hypothesized to follow a mechanism of four combined reactions —two parallel oxidation reactions which are kinetically controlled, accompanied by two series reaction where oxidation products (carbon dioxide and carbon monoxide) are converted into methane and hydrogen. Integrating this process into spaceflight missions would greatly reduce the onboard mass requirements, as these products can be used as propellant fuel. Further, under certain conditions the process could be also be adapted to produce oxygen and water, two essential life support species for space missions.

This paper explores scale-up alternatives to implement low-temperature continuous gasification processes as potential waste management processes. A kinetic model formulated from batch gasification experiments is integrated into a process simulator along with a detailed thermodynamic description.

Emphasis is placed on the preparatory treatment of the waste slurries, the reactor module, and equilibrium stages for recycling reactants and recovering the catalyst. Detailed characterization of low and mid fidelity simulants is presented.

(534f) Performance Evaluation of Accelerated Carbonation of Basic Oxygen Furnace Slag Via a Rotating Packed Bed: Modeling, Analysis and Maximization

Pan, S. Y., National Taiwan University
Chiang, P. C., National Taiwan University
Chen, Y. H., National Taipei University of Technology
Chang, E. E., Taipei Medical University
Tan, C. S., National Tsing Hua University
Originally presented on: 11/6/2013 16:55:00 - 17:15:00

Performance of accelerated carbonation by basic oxygen furnace slag (BOFS) in a rotating packed bed (RPB) was evaluated in this investigation. The design of experiments (DOE) was introduced to evaluate the effect of different operating parameters including reaction time, particle size of BOFS, temperature, liquid-to-solid (L/S) ratio, and rotating speed on carbonation conversion of BOFS. A quadratic response surface was established statistically to yield maximum carbonation conversion based on experimental data. In addition, both the fresh and carbonated BOFS were characterized by the thermogravimetry and differential scanning calorimetry (TG/DSC), X-ray diffraction (XRD), and scanning electron microscopy equipped with X-ray energy dispersive spectroscopy (SEM-XEDS) to confirm the results of accelerated carbonation of BOFS in an RPB.

Furthermore, by combining the process chemistry, reaction kinetics, and mass transfer, a prediction model for carbonation of BOFS within an RPB, with an error less than 10%, was well-developed. The enhancement factors for mass transfer coefficient in case of carbonation of BOFS within an RPB were determined accordingly. In comparison with the developed model and the RSM results, a standard operation procedure (SOP) was proposed for maximizing the carbonation conversion of BOFS. It was thus concluded that the accelerated carbonation of BOFS using an RPB is a promising process for CO_2 capture due to its higher carbonation conversion achieving in a short reaction time.

(534g) Production of Alumina From Fly Ash By a Mild and Cleaner Hydrometallurgical Process

Zhong, L., Institute of Process Engineering, Chinese Academy of Sciences
Duan, D., Institute of Process Engineering, Chinese Academy of Sciences
Zhou, E.,
Han, H.,

Originally presented on: 11/6/2013 17:15:00 - 17:35:00

Fly ash is a large volume solid waste generated during the coal combustion by coal-fired power plants, wherein the valuable alumina content could be up to 30-50%. A mild hydrometallurgical process to produce alumina from fly ash only by alkaline solution at comparatively low pressure and temperature was investigated. This novel process is also a cleaner process, where the excess alkaline solution reactant is recycled to the leaching of fly ash, followed by the crystallizing of sodium aluminate hydrate and then its seed decomposing to prepare aluminum hydroxide, calcined to produce alumina. By controlling the phase transformation of the multiphase complex components in the fly ash, overall more than 90% of alumina was recovered from fly ash. The final residues could be utilized as feedstock in construction materials. Furthermore, the reaction mechanism of the process is investigated. This mild and cleaner process could also be appropriate for the produce of alumina from other high Al-containing solid wastes.

(565a) Open-Cell Metal Foams As Enhanced Catalyst Supports for Heat Transfer Intensification in Tubular Reactors

Bianchi, E., University of Erlangen-Nurnberg
Visconti, C. G., Politecnico di Milano
Groppi, G., Politecnico di Milano
Schwieger, W., Friedrich-Alexander-University Erlangen-Nurnberg
Tronconi, E., Politecnico di Milano
Freund, H., Max Planck Institute for Dynamics of Complex Technical Systems
Originally presented on: 11/6/2013 15:15:00 - 15:36:00

In the context of process intensification, structured catalyst supports offer the advantage of a better control on the geometrical properties and consequently on the effective transport properties of the catalytic bed. In this regard, open-celled foams are of particular relevance. They represent a class of consolidated porous media which is characterized by high geometric surface areas and high void fraction [1]. Moreover, high conductive catalyst carriers can help to minimize the hot/cold spots and to prevent mechanical-strength and thermal shock limitations, thereby increasing the reactor and process efficiency. Thus, in this work the heat transport properties of metal foam supports suitable for highly exo- or endothermic catalytic gas phase reactions in tubular reactors are characterized adopting a combined experimental and modeling approach.

The experimental heat transfer data [2] was acquired with samples of different geometries and bulk solid materials in a lab-scale test rig which consists of a 10 cm long foam bed fitted in a stainless steel pipe of 28mm I.D. placed in a thermostatic chamber. Steady-state axial and radial temperature profiles were collected inside the foam beds at heating and cooling conditions and at different flow rates. Both the effects of the solid matrix (cell density, porosity, thermal conductivity) and of the saturating fluid (N2 and He) on the heat transfer have been investigated.

Afterwards, 3D representations of selected samples have been generated from the scanned images with X-ray micro computed-tomography (μCT), a non-destructive technique enabling resolutions in the order of microns, and afterwards investigated by finite volume analysis (FVA).

This detailed study points out the determinant role of the solid phase to the effective heat conductivity of the foam bed, allowing for identifying the dependency on the thermal conductivity of the bulk material, on the solid fraction and on the geometrical structure of the solid matrix. At the same heating condition, the supports with the highest thermal conductivity feature almost flat inner radial temperature profiles, with the whole T-gradient confined at the foam-tube wall interface. Because this seems to be the controlling resistance in highly conductive foam configurations, the region of coupling with the reactor tube is studied in detail [3]. The effects of the flow rate, the thermal conductivity of the saturating medium and the contact quality between the foam and the wall, have been modeled and examined in detail. Moreover, the role of the pore density as well as the solid strut thickness, which are of complementary importance to allow for a complete geometrical description of the foam, have been investigated in the simulations.

As a result of this detailed study, validated on the basis of experimental data and FVA, the dependence of the foam thermal properties on geometry and flow conditions has been clarified and correlations have been established. Such engineering correlations for the foam heat transfer processes can help to guide the design of optimized catalytic reactors. In summary, with open-cell foams as innovative catalyst supports process intensification can be achieved by taking advantage of the enhanced heat transport properties that are superior to conventional packed bed systems.

References

1. Inayat A., Schwerdtfeger J., Freund H., Körner C., Singer R. F., Schwieger W., Chem. Eng. Sci., 2011. 66(6): p. 1179-1188.

2. Bianchi E., Heidig T., Visconti C.G., Groppi G., Freund H., Tronconi E., Chem. Eng. J., 2012. 198-199: p. 512-528.

3. Bianchi E., Heidig T., Visconti C.G., Groppi G., Freund H., Tronconi E., Catal. Today, 2013, (submitted).

(565b) CFD Simulation of Heat Transfer in a Microchannel Reactor for Fischer-Tropsch Synthesis Process

Kshetrimayum, K., Seoul National University
Park, C., Seoul National University
Jung, I.,
Park, S., Seoul National University
Na, J.,
Han, C., Seoul National University
Originally presented on: 11/6/2013 15:36:00 - 15:57:00

CFD Simulation of Heat Transfer in a Microchannel Reactor for Fischer-Tropsch Synthesis Process

K. Kshetrimayum, C. Park, I. Jung, S. Park, J. Na, C. Han

School of Chemical & Biological Engineering, Seoul National University,

Seoul 151-744, Republic of Korea.

Introduction

Microreactors can have enhanced process intensification due to its reduced mass and heat transfer distances, when compared to conventional technology. In this work, heat transfer effect in a microchannel reactor adapted for Fischer-Tropsch (FT) synthesis process is being studied. The process involves substantial heat effect due to high exothermic nature of FT- reacion (heat of reaction = 165 kJ/mol of reactant CO). Computational Fluid Dyanmic (CFD) modeling and simulation of heat transfer in a microchannel can give insights on the thermal behavior of such a reactor. The knowledge of heat transfer effect can then be used to develop a design and obtain tuning parameters for a desired reactor performance. Understanding the heat effect is essential to achieving an efficient coolant system design that is considered key to the performance of microreactor for FT synthesis applications. The main objectives of the present work have been i) to study the heat transfer effect in a cross flow microchannel reactor for different types of coolants (subcooled vapor, liquid water and saturated water) and ii) to investigate the tuning parameters and best operating conditions that can maintain the reactor internals to a near isothermal condition for a desired reactor productivity.

Methodology

Reaction scheme and kinetics as proposed by Marvast et al.(2005) was considered for the present study. Heat generation profile was independently calculated by simulating the simultaneous reactions in an isothermal plug flow model using commercial software-ASPEN. The generated heat profile is implemented as source condition in the process channels of a microchanel block comprising of alternate layers of process channel planes and coolant channel planes [Arzamendi et al., (2010) for detail geometry of microchannel block considered]. In the microchannel block, heat is transferred from the process channel (where the exothermic reaction is occurring) to the adjacent coolant channels through channel walls. For this purpose, a 3D Computational Fluid Dyanmic (CFD) model of a cross flow microreactor was developed in commercial CFD software- ANSYS FLUENT.

Results

The heat transfer effect within or interior of the microchannel block was examined for various scenarios that can be adapted to the actual industrial operation of compact microchannel reactor for Fisher-Tropsch synthesis purpose.

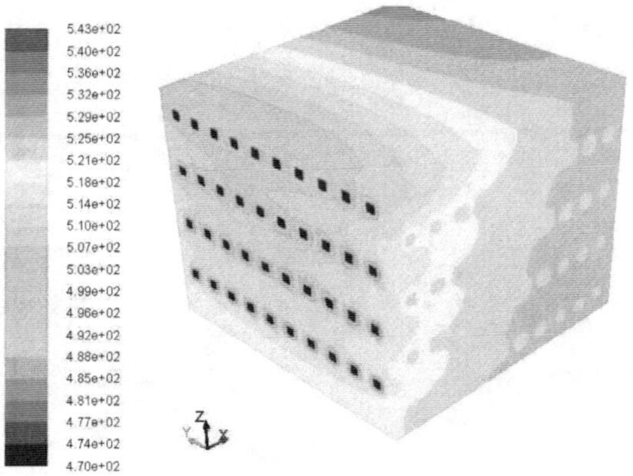

Figure 1. Temperature profile over the micro-reactor block.

Table 1: Effect of heat generation rates (Q)

	Q	100 x Q
Heat flux rates, J/s.m2	8140	13700
Syn gas exit temp, K	491	494
Coolant exit temp, K	494	497

Table 2: Effect of coolant types

	Single phase		Multiphase
	Vapor(v)	Water(l)	Sat. Water
Heat flux rates, J/s.m2	16000	13700	168200
Syn gas exit temp, K	528	494	--
Coolant exit temp, K	530	497	458

Discussion

CFD simulation of heat transfer in a microchannel with process channel as heat source(due to exothermic FT- reaction), gives insights on the flow and thermal behaviors of a microchannel reactor. Using the developed CFD model for cross flow microchannel block, effects of different heat generation rates (implying effect of different syn-gas feed rates) were studied. The results help to identify the appropiate coolant flowrate to be used for a given syn-gas feed rate. Effects of coolant type (subcooled water vapor, liquid water and saturated water) were also studied. The result indicates that saturated water as coolant can be the right choice to maintain near isothermal conditions in the reactor internal. Further, using the CFD model, tuning parameters and operating conditions for a desired reactor performance are being investigated.

References

1. M.A. Marvast, M. Sohrabi, S Zarrinpashneh, G. Baghmisheh, (2005) *J. Chem Engg.,* Vol 28.

    1. G. Arzamendi, P.M. Dieguez, M. Montes, J.A. Odriozola, E.F. Sousa-Aguiar, L.M. Gandia, (2010). *J. Chem Engg.*, Vol 160.

(565c) Research On the Enhanced Heat Transfer in Shell and Tube Heat Exchanger Based On Flow Path Optimization Idea

Xiao, X., School of Chemical Engineering and Technology
Xiao, H., Tianjin University
Zhang, L., Tianjin University
Originally presented on: 11/6/2013 15:57:00 - 16:18:00

In this paper, a novel heat transfer enhancement idea called flow path optimization idea was presented. By analytical calculation and numerical calculation, the new idea quantified the effect of flow pattern, flow path dimension, flow path division and geometric construction of element on the hydraulic performance and thermodynamics performance of the shell and tube heat exchanger. The new idea could provide heat exchanger with enhanced heat transfer elements with optimized flow path for specific case, and it also could offer simple rapid algorithm to predict the heat transfer rate and pressure drop of heat exchanger combined with multiple enhanced heat transfer elements, the elements contained the insert elements, rough surface and extended surface tubes, enhanced supporting elements (baffles), etc. Two industrial unit renovation examples, including one helical baffles heat exchanger with varied helical angles and one heat exchanger with finned tube and rod baffles were studied using the flow path optimization idea. The increased heat transfer rate were 32% and 13% respectively, the decreased pressure drop were 23% and 58% respectively, and the calculation errors were within 10%.

(565d) Pervaporation, An Efficient Tool for Separation of Azeotropic Mixtures in Ester Synthesis

Lux, S., Graz University of Technology
Winkler, T., Graz University of Technology
Siebenhofer, M., Graz University of Technology
Originally presented on: 11/6/2013 16:18:00 - 16:39:00

The bulk chemicals esters contribute to commodity chemicals in nearly unlimited variety. Though well investigated, the technology of ester formation may offer a variety of challenging characteristics regarding catalysis, operation conditions and downstream processing. Methyl acetate synthesis via esterification of acetic acid and methanol serves as a welcome model reaction, since methyl acetate forms low boiling binary azeotropes with methanol and the by-product water. Reverse hydrolysis reaction and transesterification are further examples facing the same issue.

The esterification reaction itself was catalyzed by commercial cation exchange resins. The operation temperature was kept well below boiling point. The reaction kinetics was modeled with an adsorption based approach accounting for the fact that the reactants show different adsorption behavior on the resin due to their different polarity. The reaction was performed under acetic acid limiting conditions and excess methanol and water were then separated from methyl acetate by pervaporation.

Pervaporation is a promising technology for separation of azeotropic mixtures. The main benefits are based on the moderate operation temperatures, partial evaporation and the nature of the driving force. This unit operation can handle higher water contents than reactive distillation technology. Moderate temperatures may suppress unwanted side reactions such as dehydration (ether formation) of methanol.

With membranes featuring hydrophilic surface properties methanol and water can pass the membrane, whereas methyl acetate is concentrated in the retentate phase. The binary mixture methyl acetate/methanol is the technological bottleneck of the isolation process. Appropriate membranes are needed for separation. Depending on the degree of crosslinking of the membranes and the composition of the feed, composite membranes show different swelling behavior which affects their selectivity and the permeate fluxes. Therefore membrane selection still needs experimental support and modeling of the permeate flux has to account for composition dependent swelling.

Based on the outcome of investigations pervaporation-assisted isolation of methyl acetate is an alternative route to the state of the art process.

(565e) A Short-Cut Model for Distillation in Rotating Packed Beds for Conceptional Process Design

Sudhoff, D., TU Dortmund University
Leimbrink, M., TU Dortmund University
Lutze, P., TU Dortmund University
Originally presented on: 11/6/2013 16:39:00 - 17:00:00

A Short-Cut Model for Distillation in Rotating Packed Beds for Conceptional Process Design

Authors: Daniel Sudhoff*, Mathias Leimbrink, Philip Lutze

Keywords:

Rotating Packed Bed, Distillation, centrifugally-enhanced separation, short-cut model

* Dipl.-Ing. Daniel Sudhoff

TU Dortmund University

Phone: +49 (0) 231/ 755 6002

Email: Daniel.sudhoff@bci.tu-dortmund.de

Web: www.fvt.bci.tu-dortmund.de

To meet future challenges in the chemical and bio-chemical industries additional requirements for chemical production plants arise. That includes, besides others, quick market entry by saving time and cost to operation, increased flexibility with respect to the capacity by the possibility to adjust process parts, increased flexibility with respect to changes in the composition of the feed and a dynamic operation by additional control levers. These requirements not only apply for reaction but also for separation processes and call for new and innovative separation apparatuses.

A potential device to meet these requirements is the rotating packed bed (RPB) that offers intense mass and heat transfer and a flexible operation by the adjustable rotational speed at a compact and space economic design. By the rotation of a cylindrically shaped packing at a large number of revolutions per minute the RPB creates high centrifugal forces in a centrifugal field. The phase contact between a liquid and a vapor occurs inside the rotating packing and therefore inside the centrifugal field. Due to the high applied forces to the liquid and the intense mixing of the vapor very high mass transfer rates can be achieved. Additionally, packings with a large surface area can be applied due to the high centrifugal forces applied to the liquid.

Both the large surface area and the intense mixing enable high separation efficiency at a small volume and therefore relatively compact devices. The additional degree of freedom, the rotational speed, which has a significant influence on the separation efficiency, gives the opportunity of a quick change in separation performance of the RPB and therefore of a quick response to changes in the feed streams or in the requirements for the product streams. These advantages make the RPB a potential candidate to meet the above mentioned requirements for chemical production plants.

Although having first been described already in the 1980s by Colin Ramshaw [1] and the RPB's potential advantages over columns RPBs have not found their way into the toolbox for the conceptional process design yet. Except for the application as a reactor for the production of nano-particles and for special absorption and stripping processes mainly in Asia, RPBs for distillation have not been studied systematically. Although approaches for the prediction of the separation efficiency of RPBs in analogy to those in packed columns have been proposed in some of these research activities, a reliable mathematical description is still missing for the process design. Due to this lack the RPB technology is often not considered in the conceptional design of new processes, although it might be advantageous for certain applications. The scope of this work is set into this context.

The aim of this investigation is to develop a short-cut model for the description of the separation efficiency of RPBs for distillation which is simple but accurate enough to a certain degree and allows quantifying the separation efficiency and the potential failure also while extrapolating. These requirements are due to the use as a model at an early stage of the conceptional process design to evaluate the feasibility of the RPB for distillation for a given problem without a large effort. Based on this a decision on whether the RPB should be investigated further for the given problem can be done. To derive the short-cut model a comprehensive overview of existing approaches for the description of the separation performance of RPBs for distillation has been created from literature. These approaches have been evaluated on the bases of general applicability for the special geometry and force fields inside an RPB, of applicability for a single or for general RPBs and sizes and of its potential for extrapolation in a given range. Additionally, the complexity of the approaches has been evaluated in order to use them as a short-cut model. Based on this evaluation a short-cut model has been derived using the reduced literature approaches with own modifications by the identification of important process and design parameters. The short-cut model has than been adapted to an RPB pilot plant facility consisting of three consecutive rotors. For this purpose finite distillation experiments varying the important operating parameters have been carried out in a pilot plant multi-stage counter-current rotating packed bed (MSCC-RPB) that has been presented in earlier works [2]. Doing so the short-cut model has been tested and validated against own experimental data. The developed short-cut model fulfills the requirements for the conceptional process design and enables the process engineer to consider RPBs for distillation in an early process design stage.

[1] Ramshaw, C.: "Higee" Distillation – An Example of Process Intensification, Chem. Eng. 1983, 13, p. 13-14.

[2] Sudhoff, D.; Lutze, P.; Górak, A.: Multi-Stage Counter-Current Rotating Packed Bed for Distillation; presentation at AIChE Spring Meeting 2013, San Antonio, USA.

(565f) Optimal Design of Reactive Distillation Coupled With Side Reactor for Tert-Butyl Acetate Production

Chen, M.,

Li, C.,

Tang, J., Nanjing University of Technology

Chen, X., Nanjing University of Technology

Fei, Z., Nanjing University of Chemistry

Cui, M., Nanjing University of Technology

Qiao, X., Nanjing University of Technology

Originally presented on: 11/6/2013 17:00:00 - 17:21:00

Optimal design of reactive distillation coupled with side reactor for tert-butyl acetate production

Ming CHEN, Chong LI, Jihai TANG, Xian CHEN, Zhaoyang FEI, Mifen CUI, Xu QIAO

State Key Laboratory of Materials-Oriented Chemical Engineering, College of Chemistry and Chemical Engineering, Nanjing University of Technology, Nanjing, 210009, PR China

Key words£ºreactive distillation coupled with side reactors; tert-butyl acetate; optimal design

Reactive distillation (RD), combining reaction and separation in a single unit, is not effective in many chemical systems because of the conflicting requirements of reaction and separation [1-3], such as the optimum conditions of temperature and pressure for distillation may be far from optimal for reaction, and the requirements of high liquid or catalyst holdup are not in consistance with the requirement of good in situ separation. Recently, a new reactive distillation integrating concept, that is a distillation column coupled with side reactors (SRC), has been proposed [4]. Since the reaction and separation are taken in different space, it maintains the benefits of in situ separation with reaction and overcomes these problems of RD column.[1, 5]

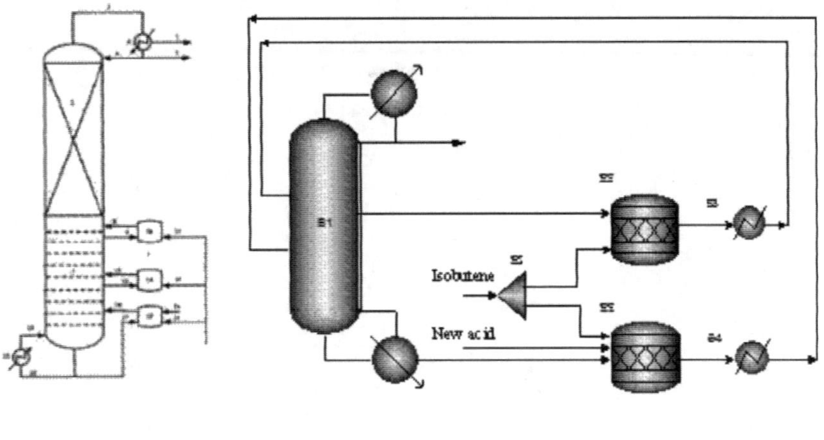

(a) (b)

Fig. 1 Schematics of a distillation column coupled with side reactors (SRC)

In this work, the SRC process for the synthesis of producing tert-butyl acetate (TBA) from acetic acid and isobutene, which is a parallel reversible reaction, was investigated. The configuration of a distillation column coupled with side reactors for tert-butyl acetate production is displayed in Fig.1. Figl (b) is the simulation flowsheet in Aspen Plus, in which B1 (Radfrac) represents the distillation column. B5 and B6 (RStoic) represent the side reactors. B3 and B4 (Heat) are heat exchangers. B2 (Split) is used as a splitter to distribute the feedstock isobutene in different side reactors. The entire liquid stream leaving the trap-out trays is completely rerouted through a reactor before it is fed back to the column. The make-up acetic acid is fed to the nethermost reactor, and the isobutene is distributed by splitter to side reactors.

The effects of the various operation parameters and configuration parameters on the performance of expenditure-to-revenue ratio (ERR) are well studied, to preliminarily explore the interaction between reaction and distillation. As the feed rate of acetic acid increases, the total cost decreases first and then increases, while the revenue is monotonically increasing (Fig.2). Thus an optimal feed rate is recommended leading to the minimum ERR. Fig.3 shows the effect of boil-up rate and bottom rate on optimum ERR. The optimum ERR decreases with the raising of boil-up rate and bottom rate. As can be seen from Fig. 4, the optimum feed rate, namely the reaction capacity, becomes lager as the number of the side reactors increases. It can also be observed that the optimum feed rate raise slower while RN is lager than 2, indicating that the number of side reactor 2 is proposed.

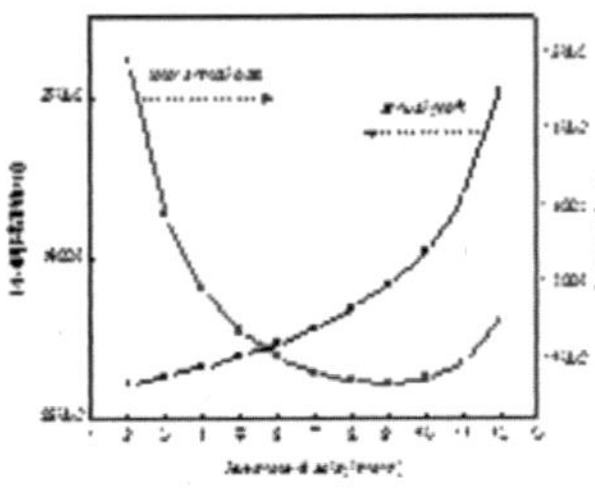

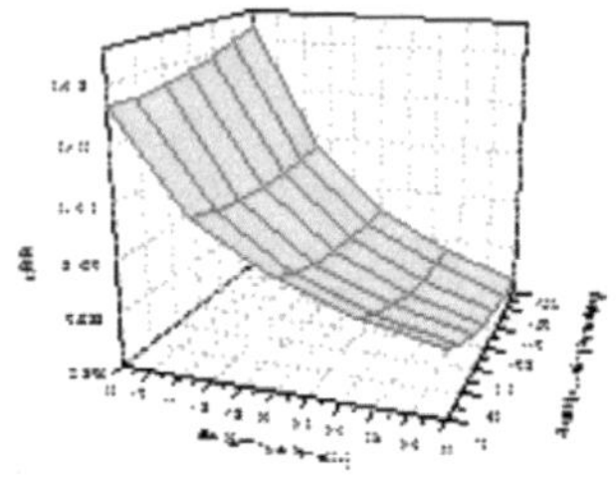

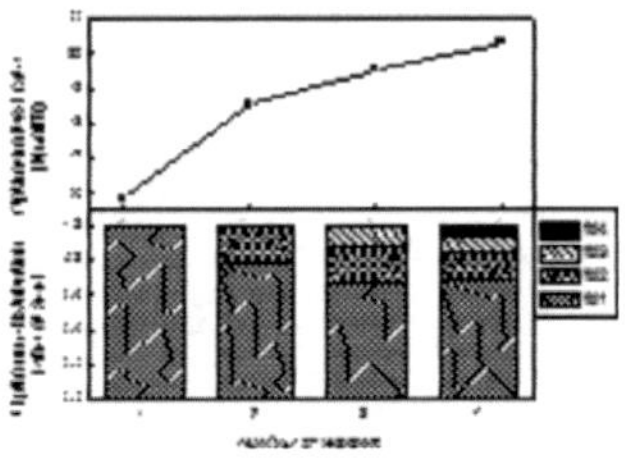

Fig. 2 The effect of feed rate on optimum feed rate and (Number of total trays=17, NR=15, RN=1)

Fig. 3 The effect of boil-up rate and bottom rate (N=17, NR=15, RN=1)

Fig. 4 The effect of RN distribution (NR=17, NRS=1)

References:

[1] Ding L H, Tang J H, Cui M F, et al. Optimum Design and Analysis Based on Independent Reaction Amount for Distillation Column with Side Reactors: Production of Benzyl Chloride [J]. Industrial & Engineering Chemistry Research. 2011, 50(19): 11143-11152.

[2] Kaymak D B, Luyben W L. Optimum design of a column/side reactor process [J]. Industrial & Engineering Chemistry Research. 2007, 46(15): 5175-5185.

[3] Harmsen J. Process intensification in the petrochemicals industry: Drivers and hurdles for commercial implementation [J]. Chemical Engineering and Processing. 2010, 49(1):

70-73.

[4] Kaymak D B, Luyben W L.
Design of distillation columns with external side reactors [J]. Industrial & Engineering Chemistry Research. 2004, 43(25): 8049-8056.

[5] Jiao Y, Wang S J, Huang K J,
et al. Design and Analysis of Internally Heat-Integrated Reactive Distillation Processes[J]. Industrial & Engineering Chemistry Research. 2012, 51(10): 4002-4016.

(565g) Hydrodynamics and Mixing Characterization in a High Shear Mixer

Vashisth, V., Indian Institute of Technology Roorkee
Kumar, D. V., IIT Roorkee
Mishra, I. M., Indian Institute of Technology Roorkee
Originally presented on: 11/6/2013 17:21:00 - 17:42:00

High shear mixers (HSMs) are widely used in process industries (e.g. food industry, paint industry, pharmaceuticals, etc.) for carrying out reactive mixing and dispersed phase size reduction. HSMs are also knows as rotor-stator-mixers and are characterized by their highly localized energy dissipation and drop size distributions. In the present work the computational fluid dynamic modeling of an existing HSM and a proposed novel HSM design is carried out using k-ε turbulence model. The existing HSM considered in the present work is comprised of square holes in the stator and a rotor with 4 blades. The proposed mixer design is comprised of tilted jets/holes in the stator, for the deep penetration in the mixing vessel. The mixing characteristics and flow patterns in both mixers are analyzed in the bulk, stator and rotor regions. The results are also reported in terms of energy dissipation and power number, as both are characteristic parameters for motor selection in the mixer design, optimization and scale-up. The simulations are carried out using transient sliding mesh with k-ε turbulence model at different rotor speeds (2000 rpm, 4000 rpm and 6000 rpm). The simulated results for power number and flow rate are compared with the existing data available in the literature and showed a variation of 8.3% and 0.8%, respectively. The proposed novel mixer has a very high efficiency as compared to the existing mixer. The mixing behavior has also been studied at different rotor speeds. Higher and uniform mixing is obtained in the proposed mixer as compared to the existing rotor stator mixer.

(566a) Greenscope Tool for Process Design and Technological Evaluation

Ruiz-Mercado, G. J., U.S. Environmental Protection Agency
Smith, R. L., U.S. Environmental Protection Agency
Gonzalez, M. A., U.S. Environmental Protection Agency
Originally presented on: 11/6/2013 15:15:00 - 15:40:00

The chemical industry is one of the most important business sectors having a relevant role for the global economy and society's quality of life. However, these benefits can involve human health and environmental costs. These challenges and improvement opportunities require the integration of sustainable development approaches to achieve quality of life enhancement without affecting the availability of ecological goods and services for new and existing chemical manufacturing processes. Therefore, the U.S. Environmental Protection Agency has created the GREENSCOPE tool to guide decision-makers, designers, and stakeholders in developing more sustainable processes.

Process design for sustainability requires the implementation of quantitative evaluation criteria that can describe if a particular process is sustainable and how sustainable it is. These sustainability criteria should identify process areas in need of redesign and key decision aspects in order to obtain an optimal tradeoff and net overall improvements. GREENSCOPE provides this quantitative assessment of sustainability by employing a set of indicators capable of transmitting and translating process performance, feedstocks, utilities, equipment, and output information into a sustainability measurement scale. These process characteristics (indicators) are categorized in the areas of environment, efficiency (material), energy, and economics, known as the 4 E's.

GREENSCOPE is presented in conjunction with a sustainability performance assessment for the manufacture of biodiesel. A set of distinct continuous processes for biodiesel production were designed and simulated. A sustainability assessment of these processes was conducted to evaluate their technical benefits and to describe how well the processes under study make use of mass and energy goods and services to manufacture a valuable product, achieve social and environmental needs, while maximizing their economic outcomes. Therefore, GREENSCOPE is proposed as a proficient tool for evaluating and designing various technologies for the production of a valuable good and identifying the most sustainable alternative.

(566b) Emission Source Characterization During An Ethylene Plant Shutdown

Wang, Z., Lamar University
Xu, Q., Lamar University
Ho, T. C., Lamar University
Originally presented on: 11/6/2013 15:40:00 - 16:05:00

Ethylene plant shutdowns not only generate huge amounts of emissions (VOC), but also waste tremendous raw material that could be utilized to generate more products. A well-planned shutdown would reduce the risks on safety, reliability, business and environment. However, fundamental and quantitative studies on shutdown flaring emission are still lacking. During an ethylene plant shutdown, plant usually cut off feed and flared off all off-spec products until the plant was de-inventoried. It would be executed following three steps: (i) Preparation of shutdown. Feed rates are reduced to minimize the liquid inventories in facilities ahead of the shutting down. (ii)Shutting down. The plant is shut down, beginning with shutting down Cracked Gas Compressor (CGC), followed by gradually cutting off the furnace feed. (iii) Decommissioning including liquid discharge, vapor discharge and N_2 purge. Almost all of the emissions come from shutting down and decommissioning.

In this paper, plant-wide dynamic simulations are employed to model the preparation of shutdown to get initial status for shutting down and decommissioning. The emission during shutting down has been counted. The decommissioning which generates most of flare source is then investigated to quantify and characterize flaring emission source during an ethylene plant shutdown. Liquid discharge is calculated by governing equation. Dynamic simulations for vapor discharge and N_2 purge are built based on isolated sections such as CGC. Deep insights of the emission source distribution and dynamic emission profiles are provided. The durations of decommissioning events calculated in this paper will help decision makings on planning plant shutdowns. The study enriches emission inventories with details for industry point sources, which have no accurate resolution on shutdowns. It also provides detailed technical support for both the industry and environmental agencies on evaluating and developing cost-effective flare minimization control strategies in the future.

(566c) Optimizing Generator Efficiency Using An Optimized Network of Loads

Fekri Moghadam, M., University of British Columbia
Vaahedi, E.,
Dunford, W. G., University of British Columbia
Metcalfe, M.,
Originally presented on: 11/6/2013 16:05:00 - 16:30:00

The existing electricity industry is unidirectional in nature. It converts only one-third of fuel energy into electricity and almost 8% of its output is lost along its transmission lines, while 20% of its generation capacity exists to meet peak demand only (i.e. in use only 5% of the time) [1].

Every electrical utility uses different generating resources to meet the total demand. As competition in the utility industry increases, each utility attempts to maximize the value of their resources at different levels of operation. The hierarchical approach for maximizing the benefits of resources is divided into several operation planning models with different time spans and computational burdens.

- Strategic long-term operation planning which covers 1 to 4 years with a monthly time step
- Medium-term optimization over a daily or weekly time steps for up to a year
- Short-term optimization with hourly or sub-hourly time steps

Permanent balance between generation and demand in real time is the key to stable and reliable operation of the electric power system. Electric demand follows seasonal and daily cycles which are predictable using demand forecasting methods. But the very short term and real-time component of system demand has stochastic characteristics imposing mismatches between supply and demand. In order to keep these mismatches within permitted levels while reducing balancing costs, a hierarchy of different balancing tasks is designed in the Energy Management Systems (EMS) of operation centres. Generation resources are committed and dispatched economically to meet demand. Since demand is a stochastic fluctuating quantity, flexible generation units have traditionally been used as balancing tools to follow the demand. As renewable energy penetration increases, the variable and uncertain characteristics of demand become even more severe. This intermittency increases the need for installing new spinning and regulation reserves and sub-hourly load following capacity. Each generation unit has an efficiency characteristic relating the input mechanical power to the output electrical power supplied to the grid. This efficiency characteristic is the most important factor used in generation scheduling and dispatch to maximize the total efficiency of the utilized resources. Meeting the reserve constraints and following the fluctuating demand trends lead to loading of the units in sub optimal operating levels.

The emergence of new metering technologies and advances in communications and control systems has made DSM a feasible and economic tool to operate smart power grids. The Smart Grid is expected to overcome the serious existing challenges such as the low efficiency of generation and a high carbon footprint [2]. Maximizing asset utilization and high penetration of renewable energy resources and Energy Storage Systems (ESS) are also targeted in the smart grid paradigm [3].

Demand Response (DR) refers to the responsiveness of power consumers to price or control signals from utility. It can be exercised in contingency events as a reliability resource or during price spikes as efficiency resource. The Federal Energy Regulatory Commission (FERC) order 719 [4] and 745 [5] promote participation in DR programs and discuss compensation rules and initiatives. Current DR programs in electricity markets are based on voluntarily or directly controlled load curtailments or load reductions to support the power grid while providing a revenue stream for participating customers and aggregators [3]. DR provides excellent opportunities to avoid/defer investments required to meet the current fluctuating demand and reduce the overall energy price. However, load curtailments or load reductions cause temporary discomfort for the participants. This inconvenience limits their deployment frequency and duration.

Energy Storage Systems (ESS) such as batteries, super capacitors, flywheels and superconducting magnetic energy storage (SMES) systems etc. have been suggested for facilitating intermittent generation integration, improving grid reliability and power quality, enabling energy arbitrage, transmission distribution investment deference and providing ancillary services in [6], [7], [8] and [9]. The main drawback of using these storage systems with a power electronics interface is their high capital cost, low energy density, low efficiency and potential adverse environmental effects [8]

Inherent storage and flexibility of some residential, commercial and industrial loads make them valuable resources for power system operators dealing with active power manipulation. The capability to aggregate and continuously control individual responsive loads based on power system requirements is called Demand Management. Some industrial processes e.g. water pumping in water storage systems, aeration in wastewater treatment plants, industrial heating and refrigeration have an inherent flexibility within their existing assets [10]. This flexibility provides required storage to be used in active power manipulation tasks such as Frequency Regulation or so called Automatic Generation Control (AGC). These processes have enough power capacity, energy storage and no adverse reactive effect on the grid that make them suitable resources for grid balance applications. Their cycle efficiency is also very close to 100%. These loads can be used to provide real-time services to power system while fulfilling their primary obligation.

This paper deals with using flexibility of distributed responsive loads (DRLs) to minimize efficiency loss in a hydroelectric power plant. Flexibility of DRLs is modeled as regulation capacity to be used in grid balance service. DRLs are dispatched along with hydroelectric plants to increase generation efficiency without compromising their primary objectives. The objective function to be minimized is generation loss in the hydroelectric power plants. Minimization of generation loss is equivalent to maximization of generation efficiency. DRLs are treated as virtual power plants (VPP) and are distributed throughout the transmission system. Characteristics of power plants, DRLs and power system are modeled as constraints in optimization problem. These constraints are as follows:

- Power plant constraints: minimum and maximum power rating, ramp rate
- Responsive loads' constraints: minimum and maximum power rating, storage and ramp rate
- System constraints: generation/demand balance

Two case studies are done to evaluate the effect of adding DRLs to regulation assets. The first case is dispatching DRLs along with all hydroelectric power plants in the system. In doing so, the operating points of the hydroelectric units move to the most efficient points on their efficiency curves. In the second case, only the swing plant which is operated in AGC mode is considered in objective function.

Results show the effectiveness of the proposed methodology in improving generation efficiency. The studied power system has 3 multiunit power plants. Daily "average produced energy to discharged water" is used as comparison quantity. The first case showed up to 2.25% of savings for 150MW of up and down regulation. The savings in second case is up to 0.2% since only one plant is considered. Sensitivity analysis shows that most of marginal savings are in the lower penetration of responsive loads.

[1] H. Farhangi, "The path of smart grid," *IEEE Power Energy Mag.,* vol. 8, no. 1, pp. 18-28, Jan.- Feb. 2010.

[2] K. Moslehi and R. Kumar, "A reliability perspective of the smart grid," *IEEE Trans. Smart Grid,* vol. 1, no. 1, pp. 57-64, June 2010.

[3] F. Rahimi and A. Ipakchi, "Demand response as a market resource under the smart grid paradigm," *IEEE Trans. Smart Grid,* vol. 1, no. 1, pp. 82-88, June 2010.

[4] Federal Energy Regulatory Commision, FERC Order 719, [Online]. Available: http://www.ferc.gov/whats-new/comm-meet/2008/101608/E-1.pdf.

[5] Federal Energy Regulatory Commission, FERC order 745, [Online]. Available: http://www.ferc.gov/EventCalendar/Files/20110315105757-RM10-17-000.pdf.

[6] G. D. Rodriguez, "A utility perspective of the role of energy storage in the smart grid," in *IEEE Power and Energy Society General Meeting*, Minneapolis, MN, 2010.

[7] A. Mohd, E. Ortjohann, A. Schmelter and D. Morton, "Challanges in integrating distributed energy storage systems into future smart grid," in *IEEE International Symp. on Industrial Electronics*, Cambridge, UK, 2008.

[8] Bo Yang, Y. Makarov, J. Desteese, V. Viswanathan, P. Nyeng, B. McManus and J. Pease, "On the use of energy storage technologies for regulation services in electric power systems with significant penetration of wind energy," in *5th Int.l Conf. European Electricity Market*, Lisboa, Portugal, 2008.

[9] R. Walawalkar, J. Apt and R. Mancini, "Economics of electric energy storage arbitrage and regulation in New York," *Energy Policy,* vol. 35, pp. 2558-2568, April 2007.

[10] I. Stadler, "Power grid balancing of energy systems with high renewable energy penetration by demand response," *Utilities Policy,* vol. 16, no. 2, pp. 90-98, June 2008.

(566d) New Process Development for Precious Metal Recycling and Premier Organosilicon Product Manufacturing

Bai, H., Momentive Performance Materials
Bahr, S.,
Originally presented on: 11/6/2013 16:30:00 - 16:55:00

Organosilicon materials are used as additives in a wide variety of products and generally provide or enhance specific attributes. Organosilicon products can serve as surface-tension lowering agents, stabilizers for polyurethane foams, additives for coating applications, antifoams, emulsifiers, sealants, etc. One of the key reactions for making organofunctional silicon-containing products is hydrosilylation – the reaction of a silanic-hydrogen functional compound and an olefin in the presence of a precious metal catalyst.

Currently, Speier's chloroplatinic acid and Karstedt's platinum-siloxane complex are commonly used homogeneous hydrosilylation catalysts for the preparation and commercial manufacturing of organosilicon compounds. These platinum compounds are soluble either in the organosilicon product or the distillation heavy waste stream. This results in significant precious metal waste, undesirable high product color, and black particle formation due to precipitation of reduced Pt. Resource limitations, increased demand and the high cost of precious metals (e.g., platinum) have led to focus on recovery of platinum from organosilicon hydrosilylation streams as an important source of cost savings in the chemical industry that also improve product quality.

In this paper, we will share two new processes developed at Momentive which have proved to be successful and cost-effective for precious metal recycling and the manufacture of premier organosilicon product. By applying these new processes, the resulting hydrosilylation products showed very low precious metal content, reduced color and virtually no black particulate formation. Thus, the quality of the resulting organosilicon products has been significantly improved. In addition, the precious metals used during hydrosilylation reactions can be recovered as elemental metal.

Momentive's new process for precious metal recovery can be described as an in-situ fixed-bed adsorption technology [a]. In this process, fixed-bed columns are designed to treat product streams directly between a continuous organosilicon production unit and product storage containers or directly between a continuous distillation unit and distillation heavy waste containers. As a result, several significant process advantages have been recognized: (1) the precious metal removal is in-situ and continuous (no extra operational cost); (2) process streams are already at desirable high temperatures before entering the precious metal recovery unit (no additional heating requirement); (3) only minor modifications of the existing facilities are required (in-situ process); and (4) highly efficient precious metal recovery (> 90%) and desirable precious metal loading (> 2%) on scavengers have been achieved. These results have been demonstrated both in lab and commercial operations.

Another new process is the use of heterogeneous precious metal catalysts (e.g., Pt/Alumina, Pt/Silica, etc.) to manufacture platinum- and color-free organosilicon products via hydrosilylation. Use of heterogeneous precious metal catalysts allows hydrosilylation reactions to be carried out either in batch or fixed-bed reactor (FBR) configurations. Compared to a batch-wise process, a continuous FBR process shows several significant advantages: (1) production is continuous (lower operational cost); (2) based on the high space-time-yield achieved in lab demonstration, high product output is possible at relatively small commercial-scale FBR; (3) potential for reduced footprint and investment; and (4) easy recovery of precious metals since they are retained in the FBR.

At the end of both processes, precious metals can be refined and recovered as elemental metal.

[a] H. Bai, "In situ platinum recovery and color removal from organosilicon streams", Ind. Eng. Chem. Res., 51, 16457-16466, 2012.

(566e) The Environmental and Economic Benefits of Electronic Process Control On Chemical and Petroleum Industries

Lee, D. Y.,

Originally presented on: 11/6/2013 16:55:00 - 17:20:00

The environmental and economic benefits of electronic process control on chemical and petroleum industries

Do Yong Lee[1], Yuan Chang[1], Eric Masanet[1,2]

[1]Northwestern University, Department of Chemical and Biological Engineering, 2145 Sheridan Road, Evanston, Illinois, USA, 60208

[2]Northwestern University, Department of Mechanical Engineering, 2145 Sheridan Road, Evanston, Illinois, USA, 60208

Abstract

Process control is one of the most important factors in the chemical and petroleum refining industries for optimizing economic, environmental, and safety performance. Moreover, process control is critical because it helps to produce high value-added products using more flexible and complicated processes. The vast majority of process control systems rely on electronic components and logic.

The negative impacts of electronics, such high life-cycle energy and greenhouse gas (GHG) emissions intensities, have been studied widely and receive much attention in the research and policy communities. However, electronic process controls can also bring environmental benefit through improved process productivity, improved energy and resource efficiencies, and reduced downtime. Major chemicals processing systems that benefit include combustion systems where controls improve heat generation and reduces emissions from petroleum refineries, thermal system where controls optimize the operation temperature for the process such as a distillation, and motor systems where controls adapts appropriate motor speed to save energy, among other examples.

This study aims to quantify both the environmental costs and benefits of electronic control systems on the US chemical and petroleum process industries. The energy savings and emission reductions associated with electronic control system are calculated based on database unique database of major industrial plants in the United States, which was compiled based on hundreds of large plant energy audits under the US Department of Energy's Save Energy Now program from 2007-2011[1]. This study further considers regional impacts, whereby foreign countries could have negative impacts (e.g., where electronics are manufactured and recycled) while the US benefits from reduced energy use, emissions, and human health impacts via control system applications. The model and results can be used by manufacturers and policy makers to better understand the life-cycle environmental costa and benefits of electronics controls in US industry, and to incentivize greener procurement practices for electronics controls to minimize localized impacts overseas while maximizing environmental and economic benefits to the nation.

References:

[1]See Save Energy Now: Energy Assessments http://www1.eere.energy.gov/industry/saveenergynow

(566f) Gas Expanded Liquids for Tailoring Chemical and Physical Properties

Urena-Benavides, E. E., Georgia Institute of Technology
Medina-Ramos, W., Georgia Institute of Technology
Switzer, J. R., Georgia Institute of Technology
Ethier, A., Georgia Institute of Technology
Woodham, W. H., Georgia Institute of Technology
Senter, C., Georgia Institute of Technology
Pollet, P., Georgia Institute of Technology
Fisk, J. S., The Dow Chemical Company
Holden, B. S., The Dow Chemical Company
Liotta, C. L., Georgia Institute of Technology
Eckert, C. A., Georgia Institute of Technology
Originally presented on: 11/6/2013 17:20:00 - 17:45:00

Gas expanded liquids (GXL) are a class of solvents that combine the properties of common organic solvents with supercritical fluids. They arise from dissolution of a gas, most often CO_2, into an organic solvent, causing a volume expansion of the system. The physical and chemical properties of the solvents can be controlled by modifying processing parameters such as pressure and temperature. Local composition enhancements have shown to be important when tuning the solvent properties. This tunabilty can be exploited for carrying reactions under controlled chemical environments while facilitating separations for post-reaction processing. Additionally employing CO_2 involves a series of environmental advantages such as low toxicity, reduced flammability, high availability and recyclability. At the same time, the amount of organic solvent used is significantly reduced. GXLs technologies can be utilized for sustainable processes that take advantage of *in situ*, reversible protection and/or activation strategies. An example of activation is the reversible formation of alkylcarbonic acid from alcohols and CO_2. Alkylcarbonic acids can be used as reversible, or self-neutralizing, acids for acid-catalyzed processes for applications including the formation of ketals or diazonium intermediates toward the synthesis of dyes, and the hydrolysis of *b*-pinene. Additionally, reversible activation and protection can synergistically contribute to higher reaction performances, easy separation and recycling strategies. This presentation will highlight the benefits of employing GXL for chemical processing by tuning the chemical environment of the solvent system. Specific examples of *in situ* protection and substrate activation will be discussed, coupling efficient reaction with facile down-stream processing. The results will be analyzed in light of specific interactions occurring in GXL systems for specific reactions of interest.

(596a) Application of Intensified Equipment in Modular Processes

Bramsiepe, C., TU Dortmund
Seifert, T., TU Dortmund
Sievers, S., Bayer Technology Services GmbH
Stenger, F.,
Hamers, B.,
Schembecker, G., TU Dortmund
Originally presented on: 11/7/2013 8:30:00 - 8:51:00

In the recent years markets for chemical and pharmaceutical products have become more and more volatile. At the same time raw material prices have been increasing so that flexible and efficient production processes have become strictly required. Here process intensification can offer opportunities for a more efficient utilization of raw materials. However process intensified technologies are often limited in throughput. Especially when parallelization or numbering up is required to prepare technically relevant production capacities, loss of economies of scale can compensate advantages of improved operating cost.

Modular equipment prepares the opportunity to build processes quickly and in multiple small scale units. This helps reducing lead times which can substantially improve Net Present Value (NPV). As could be shown in [1,2], a lead time reduction from three years in the reference scenario to one year in the modular scenario can result in an NPV improvement of more than 25 percent. Moreover does the application of modular equipment allow for precise adaption of production capacity to market development. A new production unit will only be installed given suitable market development. As could be shown in [3], this approach can make small scale modular production competitive to large scale conventional production even if benefits from process intensification are not taken into consideration. Additionally can the break-even point be shifted severely. Depending on the scenario under consideration a reduction from six years to approximately one year could be achieved. This helps reducing investment risks. Combining the advantages of intensified and modular production technologies thus seems a promising approach. Nevertheless, whether such a production technology can prepare a significant economic advantage over conventional technologies does not only depend on investment and operating costs but also on a suitable market environment. Consequently the full benefit of this approach can only be revealed if production costs, investment costs and market development are investigated simultaneously.

However a combined calculation of investment and production costs together with an investigation of market development is not yet state of the art. To close this gap a tool has been developed at the Laboratory of Plant and Process Design at TU Dortmund within the F3 Factory project, a transnational European research project. For early investigations short cut simulation models are prepared within the tool. If more detailed information is available and more complex calculations are required an interface to external programs, e.g. Aspen Plus®, is available. Based on process simulation results an estimation of equipment dimensions can be performed followed by a calculation of both investment and operating costs including required personnel. A market scenario can be defined and the economic behavior of the process under investigation can be analyzed. Performing all these investigations within one single tool, allows for statistical analyses applying screening methods like Morris-one-at-a-time method [4]. With this approach parameters with great impact on the calculation result can be identified. This can help process developers identify parameters crucial for economic success and thus reduce development time and costs.

In the presentation the tool and its application will be presented using hydroformylation of propylene as a process example. The state-of-the-art process will be analyzed and drivers for the application of intensified modular production technologies will be demonstrated using NPV calculation. An innovative production technology will be introduced that uses a mass transfer intensified reaction system and cold separation technology. Economic benefits in terms of investment and operating costs will be shown and the ability to adapt production capacities to market requirements will be demonstrated. Both production approaches will be investigated in several market environments. The impact of process intensification and modularity will be demonstrated separately. It will be shown how statistical investigations can serve for technology comparison and thus support target-oriented process development.

Acknowledgement:
The research leading to these results has received funding from the European Community's 7th Framework Programme under grant agreement n° 228867.

References:

[1] C. Bramsiepe et al., Chemical Engineering and Processing 51 (2012) 32– 52
[2] T. Seifert et al., Chemical Engineering and Processing 52 (2012) 140– 150
[3] S.Lier, M. Gruenewld, Chemical Engineering and Technology 34 (2011) 809–816
[4] Morris, M. D., Factorial Sampling Plans for Preliminary Computational Experiments, TECHNOMETRIX, pp. 161-174, 1991.

(596b) Energy-Saving Design of a Dividing-Wall Heterogeneous Azeotropic Column for the Separation of Pyridine and Water

Wu, Y. C., National Taiwan University
Lee, H. Y., National Taiwan University of Science and Technology
Chien, I. L., National Taiwan University
Originally presented on: 11/7/2013 8:51:00 - 9:12:00

Design and control of a heterogeneous azeotropic distillation system for the separation of pyridine and water have been studied in Wu and Chien (2009). By adding an entrainer (toluene) into the system, two additional azeotropes are formed. Because of the minimum-boiling heterogenous azeotrope of toluene and water, a simple two-column heterogeneous azeotropic column system is devised and can be used to other separation systems with the same residue curve map (RCM) type. In this paper, a further intensified dividing-wall column is proposed to thermally-coupled the two columns to reduce the overall reboiler duty. The design of the dividing-wall column is more difficult than a conventional two-column sequence because a decanter exists in the system and also because one of the original columns is served as pre-concentrator column for the fresh feed and also served as recovery column for the aqueous outlet.

This complex two-column heterogeneous azeotropic distillation system can be thermally-coupled into a single lower dividing-wall column. A decanter is present at the top of the dividing-wall column with the organic phase totally refluxed back to the column and the aqueous outlet stream combining with fresh feed is entered into the right-bottom section of the dividing-wall column. To clearly demonstrate the energy-saving potential of the proposed dividing-wall design, the total stages of the original two columns are preserved in the new design. Only liquid splits into left-bottom or right-bottom section of the dividing-wall column and also the feed location are optimized. The result shows that a significant 29.44% reduction of the total reboiler duty can be obtained. By further considering to change the feed location to the decanter, an additional 5.55% energy reduction can be realized.

Operation and control of this process intensified design is also investigated in this paper. Previous study (Wu, et al., 2013) showed that the control performance of an extractive dividing-wall column is hamperd because of lossing one important control degree-of-freedom. However, because of this proposed design having a dividing wall at bottom part of the column, the two important control degree-of-freedom (two reboiler duties) are still present. Dynamic simulations of the proposed control strategy under feed flow rate and feed composition disturbances will be shown in this paper.

References:

Wu, Y. C.; Chien, I. L. Design and Control of Heterogeneous Azeotropic Column System for the Separation of Pyridine and Water. Ind. Eng. Chem. Res. **2009**, 48, 10564-10576.

Wu, Y. C.; Hsu, P. H. C.; Chien, I. L. Critical Assessment of the Energy-Saving Potential of an Extractive Dividing-Wall Column. Ind. Eng. Chem. Res. **2013**, 52, 5384-5399.

(596c) Approach for the Selection of An Innovative Working Pair for Absorption Chillers and Heat Pumps

Baumgartner, A.,
Beck, M., University of Erlangen-Nuremberg
Muller, K., University of Erlangen-Nuremberg
Arlt, W., University of Erlangen-Nuremberg
Originally presented on: 11/7/2013 9:12:00 - 9:33:00

The huge demand for heating and air conditioning is provided in general under consumption of primary energy. Thermally driven heat pumps work energy efficient and use waste heat streams instead of a mechanical compressor. The performance and working range of absorption chillers and heat pumps is a strong function of the applied working pair. The present work therefore is on the quest for new working pairs which overcome the present restrictions, i.e. corrosion, crystallization, etc. and show favorable working ranges as well as high performances.

As solvent acts a new class of salts called ionic liquids. Ionic liquids are molten salts at temperatures below 100°C. As ionic liquids show an almost negligible vapor pressure the desorption step is simplified. The physical and chemical properties, especially the solution properties of ionic liquids can be tailor-made due to the almost infinite combinations of cation and anion. Ionic liquids are thus a versatile tool for heat pump applications. The refrigerant selection takes into consideration enthalpy of vaporization, heat capacity and boiling conditions as well as aspects of flammability and toxicity. Though, an appropriate working pair cannot be chosen regarding one of those parameters only. This contribution shows an effective method to select an appropriate working pair for absorption heat pumps.

The presented procedure considers in the first instance the solution properties of the chosen working pair. The characteristic diagram called "Dühring Plot" shows the possible working ranges for the working pair. Usually, the Dühring Plot is deduced from experimental data. Here, the prediction tool COSMO-RS was adapted to this task. Thermodynamic prediction models, based on quantum chemistry and statistical thermodynamics enable the user to achieve information concerning the activity coefficient merely based on the molecular structure of a molecule. Concluding from the activity coefficient the saturation pressure above the salt solution can be derived in dependence of temperature and concentration. A theoretical coefficient of performance is derived thereafter. Current working pairs achieve a 10% higher COP than common water / lithium bromide absorption chillers.

(596d) Exergy-Loss-Minimization Based Process Synthesis for Cascade Refrigerant System in Ethylene Plants

Dinh, H., Lamar University
Zhang, J., Lamar University
Xu, Q., Lamar University
Originally presented on: 11/7/2013 9:33:00 - 9:54:00

Refrigeration system holds an important role in most chemical/petrochemical processes. The traditional cascade refrigeration system (CRS) used in ethylene plants includes multiple refrigerants, each of which operates at multiple temperature and pressure levels. Each component subsystem includes a compression section, a condensing section and an evaporation section containing sub-coolers, expansion valves, evaporators and flash drums. In this study, at first, in-depth thermodynamic analysis for a single-loop refrigeration system is carried out to identify exergy loss in the heat exchanger network with non-isothermal phase changes, especially at sub-coolers and evaporators in the refrigerant system. The relationship between compressor work consumption and exergy change from one pressure level to another is also demonstrated. Based on this approach, optimizations of single-component cycles are established before they are integrated into the whole CRS model. The achieved methodology is applied in a cascade refrigeration system used in ethylene plants to explore system energy loss below ambient temperature.

The procedure of process synthesis involves three consecutive steps: Thermodynamic analysis (for both single-loop and cascade refrigeration systems), Synthesis model development and solving, and solution validation and evaluation. Rigorous steady-state model is firstly constructed and employed to generate Exergy-Temperature (B-T) chart in order to analyze their thermodynamics and energy distributions. An exergy-based mixed-integer linear programming (MINLP) model is developed to minimize the overall exergy loss of the system, where multiple refrigerants with multiple recycling loops are simultaneously addressed and all of cooling demands are fulfilled. The optimal solution is subsequently examined by conducting rigorous simulations to check their feasibility and consistency. The efficacy of the developed methodology is demonstrated by a case study of a cascade refrigeration system used in an ethylene plant that employs two refrigerants, ethylene and propylene.

(596e) Manipulating Thermodynamics to Achieve Process Intensification: A Case Study of Polymer Based Systems

Deshpande, K. T., Dow Chemical Company
Anderson, R.,
Price, E.,
Stephenson, S.,
Somasi, S., Dow Chemical Company
Dixit, R., The Dow Chemical Company
Jain, P., The Dow Chemical Company
Originally presented on: 11/7/2013 9:54:00 - 10:15:00

Manipulating thermodynamics to achieve Process Intensification: A case study of polymer based systems

Kishori Deshpande, Russell Anderson, Elizabeth Price, Serena Stephenson, Sweta Somasi, Ravi Dixit, Pradeep Jain

The Dow Chemical Company

Session: 12E06: Case Studies in Process Intensification and Microprocess Engineering

As the world's population rises and new economies emerge, society requires novel solutions to meet its most basic needs, including energy, water, housing, food, health, and transportation. However, limited natural resources entail their efficient use while simultaneously enhancing the quality of life of current and future generations. In this respect, process intensification is becoming an immensely important tool to achieve overall sustainability of the chemical process industry and the reduction of our energy footprint through controlled plant size.

Single phase solution polymerization renders itself aptly for overall energy reduction by applying process intensification principles. In particular, energy intensive solvent recovery after polymerization results in high energy foot print. Selection of appropriate solvent to lower polymer solubility and form two phases i.e. polymer phase and solvent phase provides an energy efficient alternative. The heavier polymer phase can be separated using a decantation step after polymerization to remove up to 50% of the solvent leading to lower solvent separation costs. This talk will focus on two phase dispersion polymerization for energy efficient synthesis. Specifically criteria for solvent selection and its influence on process thermodynamics along with impact of processing conditions on important polymer properties will be discussed. Further, efforts on dispersion reactor modeling will also be presented.

(596f) Energy Saving in a Crude Distillation Unit By a Retrofit Design of Heat Exchanger Networks

Fazlollahi, F., Chemical Engineering Department, Brigham Young University, Provo, UT 84602, USA
Baxter, L., Brigham Young University
Rezaei, H.,
Originally presented on: 11/7/2013 10:15:00 - 10:36:00

Pinch technology is a complete methodology derived from simple scientific principles by which it is possible to design new plants with reduced energy and capital costs as well as determining where the existing processes require modification to improve performance. In this paper, by using Pinch technology, the preheating section of Shiraz refinery distillation unit energy consumption has been studied to identify opportunities to reduce energy consumption and create minimal changes to existing heat exchanger network to reach this goal.This includes simulation of the unit by a Aspen software, extracting the required information, information processing by a Aspen software and finally correcting heat exchanger network. This study domonstrates, through making changes in the network of heat exchangers, by adding a cooler and a shell& tube heat exchanger with \$ 921,800 investment, \$ $1.611*10^6$ energy saving will be achieved annually and the investment return period is equal to 7 month (0.57 year).

(596g) Energy Improvement in VCM Plant Through Process Redesign and Pinch Technology
Baksh, F., SABIC
Originally presented on: 11/7/2013 10:36:00 - 10:57:00

Pinch Technology for Heat Exchanger Networks (HEN) is in wide use now, not only for grassroot designs but also in retrofits of existing facilities for capacity enhancement and energy efficiency improvements. Limitation in use of the Grand Composite Curves (GCC), cross-pinch heat transfer regulations and subsequent optimization techniques are their disregard or disconnect from the process side. A standard chemical plant is fairly complex in its process-process interactions and it is here where greater optimization potential rests. Thus newer techniques of optimization based on integrated GCC are being utilized, with pinch analysis being applied across distillation columns, water and refrigeration networks and even Reactor schemes.

In this case study for well known technology of Vinyl Chloride Monomer, we present the effectiveness of process based pinch. The requirements of quench cooling in many chemical processing will always appear as cross-pinch heat transfer in most HEN pinch studies, but this practice is borne out of necessity, e.g. to avoid fouling and polymerization of desired product through excessive residence time over heat exchange surfaces. Such scenarios are best optimized by cost-benefit relationships, i.e. Will the energy recovery be feasible enough to accommodate the equipment down-time required for frequent cleaning? Concept of operational changes and then pinch analysis is also evaluated, which shows the positive influence of such approach on energy optimization. By introducing changes in operational philosophy of distillation columns, steam savings of 30 ton/hr are possible, in a plant where the net import is 70 ton/hr. Further savings of 5-10 ton/hr are possible through process heat recovery. If heat is recovered instead of quenching, an additional 20 ton/hr of steam can be generated. Thus the whole Unit can become nearly self-sufficient in steam generation.

(638a) New Theoretical Methodologies for Pharmaceutical Plant Cleaning

Carr, W., Newcastle University
Martin, E., University of Newcastle upon Tyne.
Talford, M., Britest Ltd
Originally presented on: 11/7/2013 8:30:00 - 8:55:00

Plant cleaning in the pharmaceutical industry is an undervalued but critical stage of processing. Cleaning with solvents and/or surfactants is often the only method capable of removing residual particles. This is often necessary prior to further processing being undertaken in the same equipment. Regulatory bodies have stringent guidelines not only for cleaning equipment but also, as importantly, relating to carryover and/or the potential contamination of subsequent batches of other products from carried over active pharmaceutical ingredients. In a typical multipurpose pharmaceutical plant there are significant implications if cleaning is not achieved effectively and efficiently, ranging from reduced plant availability through to financial penalties and possible plant closure.

Cleaning in pharmaceutical plant is often only considered once the process chemistry has been optimised. The product is regarded as the main source of financial gain in a process, but the question that is rarely asked is - what happens if cleaning is suboptimal? Cleaning challenges can cost companies millions of dollars' as they must clean plant equipment effectively to satisfy regulatory constraints. Failure to do this right first time can result in missed stage gates with processing schedules pushed back, often with their own financial consequences. The empirical approach generally adopted for cleaning protocol development results in plant that is often over cleaned, with cost implications in terms of plant availability, cleaning agent use and disposal.

More reliable methods for plant cleaning are required and thus it is necessary to consider the basic science involved in the production processes carried out prior to defining the cleaning strategy namely (1) structural features of raw materials, intermediates, product or by-products and (2) physiochemical properties of the process materials which may affect cleanability. The research in this presentation describes a new approach to cleaning using multivariate data analysis.

The first approach utilises the fact that cleaning protocols can be determined based on the identification of chemical functional groups within pharmaceutical products. A database of pharmaceutical products was created and a multivariate analysis of the variables related to structure and composition of the functional groups was undertaken. Using principal component analysis (PCA), a set of products were identified which could potentially be cleaned utilising the same approach. This provides a means to determine how products may react to certain cleaning methods with solvents and thus it can be inferred as to how easy or challenging a product may prove to clean from plant equipment.

The second approach is based on the physicochemical properties of process materials. Using a similar PCA-based approach, physiochemical properties are identified that can potentially be used to determine how process materials can be best cleaned from plant and also which products might be difficult to clean from plant equipment.

Both techniques can be used to help determine cleaning protocols based on fundamental scientific principles. Adopting this approach ensures that the cleaning agent used is appropriate to the process chemistry and, therefore cleaning is carried out right first time and as cost effectively as possible. These approaches will enable the inference of worst-case scenario products or intermediates to clean from a processing plant prior to the start of processing. The ultimate goal is for these techniques to be used to support the design of manufacturing processes taking cleaning into account from early stages of development thereby saving time and money during the processing stages.

Beta testing is currently being carried out in collaboration with a number of companies across the pharmaceutical supply chain to validate the practical implementation of this research. Through increased knowledge and understanding of industrial process cleaning, this research will support an industry-wide move towards a target of a more sustainable future.

(638b) Integration of Pervaporation Units in the Production of APIs and Intermediates

Luciani, C., University of Maryland
Buser, J. Y., Eli Lilly and Company
Tharp, G., Eli Lilly & Co
McCulley, M., Eli Lilly & Co
Seibert, K., Eli Lilly & Company
Originally presented on: 11/7/2013 8:55:00 - 9:20:00

Pervaporation is a sophisticated separation method that uses dense polymeric and/or ceramic membranes to isolate one of more components from relatively complex mixtures. This technology offers several advantages over traditional distillation techniques over traditional distillation techniques such as better selectivity, lower energy consumption, azeotrope separation, and modular design. Moreover, pervaporation systems can be considered as green processing practices and they can be integrated to flow configurations. Since it is a diffusion-controlled process, it becomes cost-effective only when the species to be permeated are minor components in the feed. Particularly, in Pharmaceutical Industry, pervaporation systems can be used to remove undesired moisture from reactions matrixes, super-saturate solutions to control crystallizations, dehydrate streams for solvent recovery, remove volatile organic compounds from waste streams, etc.

In this work, two different pervaporation units (PVM-035 and PVM-043, Pervatech) containing ceramic membranes (ceramic-HybSi internal layer) were used to investigate dehydration of organic solvents commonly used in the pharmaceutical industry (e.g., acetone, acetonitrile, methanol, propanol, . Integration of this unit to other unit operations (reactions) was also investigated.

(638c) Injection Moulding As a One-Stop-Production to Produce Pharmaceutical Dosage Forms

Eggenreich, K.,
Schrank, S., Graz University of Technology
Koscher, G., Research Center Pharmaceutical Engineering GmbH
Treffer, D., Graz University of Technology
Roblegg, E., University of Graz
Khinast, J. G., Graz University of Technology
Originally presented on: 11/7/2013 9:20:00 - 9:45:00

Dissolving or dispersing of Active Pharmaceutical Ingredients (APIs) in a molten polymer matrix is a promising method to increase the water solubility of poorly water soluble drugs. The enhanced solubility is achieved by forming various types of solid dispersion. One technology to establish such systems is the Hot Melt Extrusion (HME). Another promising technology which combines the advantages of HME with a direct shaping process is the injection moulding (IM). Thereby the materials are molten by a single screw extruder and fed into an antechamber of the screw. In the next step, the screw stops and acts as a piston applying high pressure on the material and injects it into a mould. The mould consists of different numbers of cavities with the final geometry of the product, e.g. tablets. By defining a specific shape of the cavitie(s), a broad range of geometries can be established. Beside the extrusion, injection moulding is the state-of-the-art technology in the plastic industry. Also in the production of medical devices the IM has a wide field of use.

In our project the focus of investigation is on both the possibilities but also the limitations of IM when used for the production of pharmaceutical dosage forms. The trials were conducted on an E-Mac 50 (Engel, Austria), a small production machine. The experiments followed two different production routes. In the first route, the input materials were pellets made of different polymers (PVCL-PVAc-PEG) and an API in different ratios. They were produced in a prior HME step using a ZSK 18 Twin Screw Extruder (Coperion, Germany). In the second production route, a premix consisting of the API and the polymer was fed directly into the plastification unit of the injection moulding machine. The goal of this route is to establish a one stop production where the raw material is converted into the final product within one production step showing process periods of 2 to 3 minutes.

The obtained tablets were characterized extensively. Product parameters like release behaviour, thermogravimetric behaviour, content uniformity, level of impurities and degradation products, but also physical properties like hardness and friability or the identification of type of solid dispersion were in focus of our investigation.

References:

1. Zema L, et.al.: Injection Molding and its application to drug delivery, J. Control. Release 2012.
2. Quinten T, et.al.: Evaluation of injection moulding as a pharmaceutical technology to produce matrix tablets, European Journal of Pharmaceutics and Biopharmaceutics 71 2009 145–154
3. Breitenbach J.: Melt extrusion: from process to drug delivery technology, European Journal of Pharmaceutics and Biopharmaceutics 2002, 54, 107-117.

(638d) Scaling-Up in Supercritical Fluid Chromatography

Tarafder, A., Waters Corporation
Hudalla, C.,
Iraneta, P., Waters Corporation
Originally presented on: 11/7/2013 9:45:00 - 10:10:00

Supercritical fluid chromatography (SFC) is already recognized as the method of choice for the preparative separation of chiral molecules in pharmaceutical industries. Even for the achiral separations, especially for the separation of natural products, SFC has a huge potential to grow as the method of choice. To harness the full potential of SFC in these separations, however, we need to develop better methods for scaling-up from analytical to preparative conditions.

One of the critical procedures during the method development of a preparative separation is the scale-up procedure. Screening of competing combinations of the mobile phase compositions and the stationary phases, leading to the best separation, is time consuming and expensive to carry out directly in the preparative scale. One of the better alternatives is to first identify the optimum mobile/stationary phase combination in the analytical scale and then transfer the method to a scale where the prep separation should be carried out.

Although there are standard methods available in liquid chromatography (HPLC) for scaling-up, they are not always useful in SFC. In SFC the density of the mobile phase varies significantly across the column, which results into the variation of all the mobile phase properties. The extent and the nature of variation of density in the analytical and the preparative scale can be very different, depending on (a) the dimensions of the connecting tubes and the column, (b) the particle sizes of the stationary phases, and (c) the operating temperature, pressure and flow rate. This difference in the density profiles across the column, between the analytical and the preparative system, will lead to very different chromatography unless the scaling-up procedure is guided by a systematic approach requiring additional calculations. In this presentation we will discuss a systematic approach we are developing to scale-up operations from analytical to preparative scale in SFC.

(638e) Process Transfer and Optimization of a Commercial Roller Compaction Process Leveraging the Maximum Compressive Stress Model and Design of Experiments

Panarello, A., Merck & Co., Inc.
Patel, P., Merck & Co., Inc.
Blair, D., Merck & Co., Inc
Conway, S. L., Merck & Co., Inc.
Harrison, B., Merck Co., & Inc.
Caruso, C., Merck & Co., Inc
Kline, L. A., Merck & Co.
Originally presented on: 11/7/2013 10:10:00 - 10:35:00

In the pharmaceutical industry roller compaction is becoming more attractive as a method of choice for improving material flow behavior and content uniformity over traditional granulation processes. This work herein summarizes the scale-up and tech-transfer of a High Drug Load (HDL) and Low Drug Load (LDL) roller compaction granulation (GRN) to a commercial scale compactor. A Compressive Stresses' model was employed to scale up the LDL & HDL-GRN processes from an Alexanderwerk WP120 to a Fitzpatrick 7x10 compactor. Joint process reviews identified critical gaps and/or commercial challenges and were leveraged to define primary development objectives. During the process validation campaign, poor material flow from the In Bin Hopper (IBH) to the Initial Feed Hopper (IFH) was observed as well as sub-optimized roller compactor operation resulting in various process shut downs and subsequent deviations. A Design of Experiments (DoE) campaign was successfully employed to balance compaction efficiency and achieve the desired GRN-properties.

The Maximum Compressive Stress model was able to define the commercial operating parameters that resulted in GRNs with equivalent material properties (PSD, density, porosity, and compactability) as those manufactured at pilot scale. This enabled the team to rapidly manufacture pivotal Registration Batches at commercial scale. Material balances of the initial commercial scale batches enabled rapid identification of process risks that resulted in optimizing the process to mitigate preferential API holdup within the material transfer system. A correlation can also be made between material losses within the material transfer & compaction unit operations and composite assay. Furthermore, a joint effort was undertaken to successfully execute a commercial scale DoE campaign to optimize the compaction process to reduce equipment stress (i.e., process instability) while yielding GRN properties acceptable for manufacturing bilayer tablets. GRN properties and operating conditions were not influenced by Roll Speed, however; as expected, Roll Pressure did impact granulation properties. Characterization of the Roll Pressure operating space enabled optimization of the process to reduce equipment stress without adversely altering the GRN properties. In conclusion, the LDL and HDL-GRN roller compaction process was successfully qualified and subsequent commercial batches have demonstrated process robustness.

(638f) API Recovery From Pharmaceutical Waste Streams By High Performance Countercurrent Chromatography and Intermittent Countercurrent Extraction

Ignatova, S., Brunel University
Hewitson, P., Brunel University
Sutherland, I. A., Brunel University
Originally presented on: 11/7/2013 10:35:00 - 11:00:00

When new projects are scaled up from the research and development stage into manufacturing, recovery or disposal of waste streams becomes more and more important to minimise cost and reduce both environmental and safety impact. Modern telescopic synthesis techniques and their follow-on purification processes are not always efficient enough to prevent loss of active pharmaceutical ingredients (API) in the waste stream. There are technologies available by which components of such streams can be recovered, treated or disposed of. However, most of them are based on selective filtration and aim to recover catalysts or reduce toxicity of pharmaceutical waste streams before they are released into the general water supply or burnt.

For any waste stream purification, the following stages of process optimization should be considered: prevention and minimising of environmental impact through process design; reuse of materials with no additional processing; recycling materials after such processing if required; disposal with energy recovery and finally benign disposal.

The drive towards greener processes, process economics, technology availability, and legislation, all play a part in the selection of the best available technology for waste stream processing. One such technology is High Performance Counter Current Chromatography (HPCCC) which is being evaluated as new cost effective continuous platform technology as part of a 3 year Technology Strategy Board High Value Manufacturing initiative entitled "Scalable Technology for the Extraction of Pharmaceuticals" [1-2]. This is an industry led collaborative project involving end-users, GSK and Pfizer, a technology supply company, Dynamic Extractions, and Brunel University all actively working together to improve process efficiency and reduce both cost and waste.

HPCCC method development for waste steam purification is done at the analytical scale with subsequent scale up to the preparative 1L instrument. This demonstrates the potential of the technology for the recovery of active pharmaceutical ingredients from complex waste streams in both batch and continuous processing modes. The scalability of the HPCCC process enables optimization work at the laboratory scale to be transferred directly to the 18L pilot scale HPCCC. Different scenarios of throughput/purity/yield will be presented demonstrating the flexibility of the technology to purify the target compound at the required trial grade and avoiding expensive clean up processes. This process is therefore capable of reducing the environmental impact of pharmaceutical waste streams.

References

1. Technology Strategy Board (TSB) High Value Manufacturing Award (Grant No. TP14/HVM/6/I/BD506K) "Scalable Technology for the Extraction of Pharmaceuticals (STEP) September 2009 to August 2012

2. Sutherland, I.A., Thickitt, C., Douillet, N., Freebairn, K., Johns, D., Mountain, C., Wood, P., Edwards, N., Rooke, D., Harris, G., Keay, D., Mathews, B., Brown, R., Garrard, I., Hewitson, P., Ignatova, S. Scalable Technology for the Extraction of Pharmaceutics: Outcomes from a 3 year collaborative industry/academia research programme, *Journal of Chromatography A* **2013**, 1282, 84-94

(642a) Recovery of Phosphorus From Sewage Sludge and Sewage Sludge Ashes With the Thermo-Reductive Recophos Process

Schönberg, A., University of Leoben

Samiei, K., University of Leoben

Kern, H., University of Mining Leoben

Raupenstrauch, H., Montanuniversität Leoben

Originally presented on: 11/7/2013 8:30:00 - 8:55:00

The principle of the RecoPhos process is the fractioned extraction of phosphorus and heavy metals from sewage sludge and its ash at high temperatures under reducing conditions. The RecoPhos process transforms a problematic waste into several streams of secondary raw materials, leaving behind a minimum of solid waste. The aim of the planned RecoPhos project is to develop a completely new P-recycling process, which can avoid several of the known chemical-technical problems that are still limiting the possible success of phosphorus recycling.

<>

<>Introduction

There is no possibility to find a replacement for the resource phosphorus. It is an essential element for all life forms and its exploitable deposits are getting scarce quickly [1]. This and also the fact that phosphate mining and refining has severe bad effects on the environment, like enormous land use; and the production of vast amounts of dangerous wastes [2], will make it inevitable in the near future to recycle phosphorus from every suitable P-containing waste. Sewage sludge (coming from municipal wastewater treatment) is considered to be one of the most important alternative Phosphorus sources, as in this sludge all of the phosphorus discharged by humans is concentrated and can, if properly collected and treated, be re-used completely. Its inorganic dry matter (= ash content) has a phosphate content of 30 up to 70% of the content of phosphate rock produced from natural phosphate ores [3]. In the project, the process will be developed using modeling and lab scale experiments. Using these results, the basic design of a pilot scale plant will be carried out, a market study will be performed and the environmental impact of the process compared with the status quo will be evaluated in a Life Cycle Assessment (LCA) study [4].

<>The RecoPhos process

The core technique of the RecoPhos process, the so-called InduCarb reactor, works after the same chemical principle as the Woehler process [5], producing amongst others white phosphorus as a product.

$$Ca_3(PO_4)_2 + 3SiO_2 + 5C \rightarrow 3CaSiO_3 + 5CO + P_2$$

The above shown summary reaction runs in the inductive heated bed of carbon as illustrated schematically in figure 1.

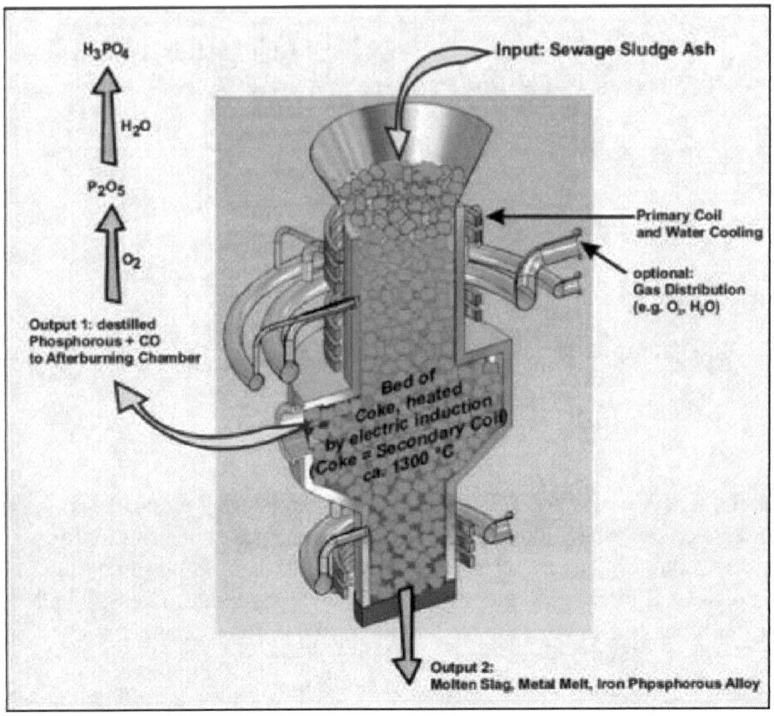

Figure SEQ "Abbildung" *Arabic 1: InduCarb reactor in the RecoPhos process

RecoPhos approaches the suitability of the following gas treatment aggregates very similar as used in waste and sludge incineration plants:

- Quench (temperature > 280 °C) for the removal of low boiling heavy metals and compounds as dust, leaving the phosphorus gaseous;
- Quench (temperature < 100°C) for the recovery of white phosphorus from the off-gas stream;
- Scrubber for the removal of acidic compounds and Hg;
- Dry and wet electrostatic precipitator for the removal of dusts and aerosols;
- Activated carbon adsorber and cloth filter or activated carbon cloth filter in appropriate positions;
- Afterburner, if the production of phosphoric acid is wished.

To be able to perform experiments, a fully operational continuous bench scale RecoPhos plant will be designed and constructed in the course of the project. This comprises the planning and basic engineering of the reactor as well as of periphery like frequency converters, off-gas cleaning and automation technology.

<>

<>Modeling and Simulation

The mathematical modeling and simulation consists of two parts: thermodynamic process simulation, and mathematical modeling of the inductive heated chemical reacting packed bed of carbon particles. The design will be constantly enhanced according to the experimental results. As a last step, using the most important results, the basic design of a pilot scale will be created.

The modeling of the InduCarb reactor also consists of two parts. At first the creation of a macroscopic model is necessary to get an overview of the occurrence in the reactor. Also it will be possible to get the dimensions for a reactor design. The aim of the model calculation is to find the necessary parameters for the dimensioning of the reactor. The correct choice of the reactor volume is decisive for the attainable conversion in the process. Furthermore statements relating to the design should be possible. The optimal height to diameter ratio of the reactor is an important statement here. Furthermore, modes of operation could be optimized in an existing reactor. Knowledge of the interactions between the processes in the reactor gains a deeper insight into the process. So you can make settings for targeted experiments.

The second part is a mesoscopic model that will be developed by MUL adapting on an existing transient, 2-dimensional heterogeneous model. The goal is to optimize the heat input, the product quality and the energy efficiency.

All over it is necessary to find out the key components and the reactions in the InduCarb reactor. Also a reducing of the surface of the carbon bed during operation has to be examined. Furthermore the input of additives should be checked (e.g. for minimizing the consumption of the reduction agent carbon or for an alternative energy supply to the reactor).

<>Summary

Recycling of Phosphorous from sewage sludge provides a sustainable solution for production of this finite resource essential for human nutrition. It will also be a solution to the environmental problems of Phosphate rich effluents from waste water treatment plants. To this end, the RecoPhos project employs the innovative InduCarb reactor design in which the ashes from sludge incineration plants will trickle down an inductively heated packed bed of Carbon. The reactions taking place inside the reactor for production of high quality Phosphorous are based on the same principle as the well-known Woehler process. The final design of the overall process will be based on the results of modeling and simulation in conjunction with experimental research at the pilot scale.

<>Acknowledgement

The RecoPhos research project has received funding from the European Union Seventh Framework Programme (FP7/2007-2013) under grant agreement no. 282856.

<>Literature

[1] Cordell, D.; Drangert, J.-O.; White, S. (2009). The story of phosphorus: Global food security and food for thought. Global Environmental Change 19: 292-305

[2] Sartorius C., von Horn J., Tettenborn F. (2011) Phosphorus recovery from Wastewater ? State-of-the-art and future potential, *International conference on Nutrient Recovery and Management 2011*, January 9-12, Miami, Florida, USA

[3] S. Donatello, D. Tong, C.R. Cheeseman. (2010). Production of technical grade phosphoric acid from incinerator sewage sludge ash (ISSA). Waste Management 30, 1634?1642

[4] RecoPhos Grand Agreement, Annex I ? Description of Work, 2011

[5] Corbridge, D.E.C. (1995). Phosphorus: An outline of its chemistry, biochemistry and uses. Elsevier. pp. 556.

(642b) Data Availability and Management for Process Sustainability Evaluation: Greenscope

Mukherjee, R., U.S. Environmental Protection Agency
Ruiz-Mercado, G. J., U.S. Environmental Protection Agency
Smith, R. L., U.S. Environmental Protection Agency
Gonzalez, M. A., U.S. Environmental Protection Agency
Originally presented on: 11/7/2013 8:55:00 - 9:20:00

Chemical products are obtained through processes that transform valuable resources into desired products, byproducts, and undesirable wastes and emissions. A particular product can be obtained from alternative processes involving varying amounts and types of resources, utilization and byproduct formation. Therefore, it is necessary to develop and identify the most sustainable process among many design alternatives.

The U.S. Environmental Protection Agency has created the GREENSCOPE tool (Ruiz-Mercado et al, 2013) (EXCEL® version) to guide decision-makers, designers, and stakeholders in developing more sustainable processes. GREENSCOPE calculates about 140 sustainability indicators and estimates chemical process sustainability in the areas of environment, efficiency, energy, and economics. This tool can be used for chemical process sustainability evaluation either in the early design or at the operational stage and aid in proper decision making to yield the most sustainable process possible.

GREENSCOPE process indicators are calculated using process data and chemical property data. The scope for successful implementation of chemical process sustainability relies on the availability and management of representative data from various sources. This includes chemical process data, classification lists, potency factors, and physicochemical, thermodynamic, and toxicological data. The complexity of data handling and management required for chemical process sustainability evaluation with GREENSCOPE has lead to the advent of a standalone software version. The standalone software tool is capable of handling challenges with data availability from various sources and heterogeneity of different chemical processes. This work describes development of the JAVA based GREENSCOPE tool for chemical process sustainability evaluation. This tool is designed to be integrated and available to other design tools such as simulator packages, optimization tools, sensitivity tools, etc.

In the GREENSCOPE JAVA version tool, a relational database for data management has been implemented to handle data from different sources. Both local and online databases are connected with GREENSCOPE. Currently, database creation, upload, modification, and access were performed using JAVA database connectivity functions. Different relational databases were created for proper handling of data using MySQL.

This presentation will describe the indicators used for sustainability evaluation of chemical processes. A detail description of different methods for obtaining required data and their handling methods for sustainability estimation with different indicators will be shown. It will also show how choices between different process design alternatives can be made using GREENSCOPE.

Ruiz-Mercado, Gerardo J., Michael A. Gonzalez, and Raymond L. Smith. "Sustainability Indicators for Chemical Processes: III. Biodiesel Case Study." *Industrial and Engineering Chemistry Research.* **DOI:** 10.1021/ie302804x

(642c) Development of a Sustainable Phase Transfer Catalysed Continuous Process for a Pharmaceutical Intermediate

Teoh, S. K., Institute of Chemical & Engineering Sciences
Sa-ei, K., Institute of Chemical & Engineering Sciences
Noorulameen, M. S.,
Ng, Y. L., 1Institute of Chemical & Engineering Sciences
Sharratt, P. N.,
Wheelhouse, K.,
Rutter, A.,
Originally presented on: 11/7/2013 9:20:00 - 9:45:00

Three of the major environmental problems in the pharmaceutical industry are solvent-laden waste, catalyst-containing waste and waste produced from side or by reactions. Batch processing dominates pharmaceutical and fine chemical manufacturing. Compared to continuous processing, batch processing is inherently less mass and energy efficient. Continuous flow also provides the possibility of continuous solvent recovery even if a more dilute reaction be required to maintain mobility.

Phase Transfer Catalysis (PTC) is a long established procedure for the facilitation of reactions between two immiscible phases using quaternary ammonium salts. From the perspective of green chemistry, PTC provides environmental benefits in eliminating, reducing or replacement of environmentally incompatible dipolar aprotic solvent such as DMF, DMSO and THF. It could replace/eliminate stronger and/or highly toxic bases (NaH, organic base), increase conversion and/or selectivity, and hence reduce waste produced. We evaluate the synergetic environmental advantages of operating PTC in a continuous flow system.

The model chemical system used in our investigation is the heteroatom O-allylation of 3-phenyl-1-propanol with allyl bromide. A typical chemical system reported in the literature using 2-phenylethanol was conducted in plant scale batch wise at a reaction temperature of 65-75°C using NaOH micropearls, toluene (solvent) and tetrabutylammonium hydrogen sulfate (TBAH) reacted for at least 16 hours to give the product as a toluene concentrate (approximately 95 mol% product by GC). From our batch process development on 3-phenyl-1-propanol system, an effective amount of PTC accelerates the reaction by about 8 fold at room temperature. In the absence of PTC, the reaction temperature has to be elevated to achieve the required reaction rate and conversion. It was found that the model reaction could be operated at room temperature with a 0.028 mol eq. of Tetra-n-butylammonium bromide (TBAB). Using NaOH solution (50% w/w) instead of NaOH pellets, the reaction could perform as well (95% conversion in 1 hour with 97% product purity by GC) without the need of toluene and this also facilitates flow processing.

The model reaction has been demonstrated at scale both batch wise (≈ 2 liter) and continuously (≈ 3 liter/hour). In the continuous process, the reaction is operated counter-currently in a liquid-liquid extractor.

The reaction is robust and takes about 100 mins to reach its steady state with a comparable reaction conversion to the batch process. The synergetic sustainability benefits of operating phase transfer catalyzed process in a continuous flow system are being compared to the traditional batch process from reaction to work-up. Sustainability metrics such as the efficiency of using chemical reagents/solvent/energy, the extent of aqueous/solids waste generation, E factor, equipment efficiency, and cost efficiency will be discussed.

(642d) Innovative Sustainable Materials & Processes

Enright, T., Xerox Research Center of Canada
Goredema, A.,
Originally presented on: 11/7/2013 9:45:00 - 10:10:00

Xerox Research Centre of Canada (XRCC) is the centre for global materials innovation in the Xerox Corporation. The mandate of XRCC is to develop all of the materials that are used in Xerox devices and to be on the forefront of materials that will enable new technologies. One of the primary goals of the centre is to deliver value to our customers through more sustainable and environmentally responsible materials and processes. This is being achieved through application of green chemistry and engineering from the conceptualization and design phase, through development and scale-up, and all the way through technology transfer to manufacturing. XRCC has commercialized a number of sustainable technologies, including functional materials that enable products to consume less energy, processes that minimize the overall amount of waste, and materials with higher amounts of bio-renewable content. A general overview of XRCC's sustainable product and process strategy will be presented, along with a highlight example that demonstrates how a multistep solvent-intensive process was redesigned to eliminate all waste solvents.

(642e) Dry Granulation of Blast Furnace Slag for Heat Recovery

Kofler, M., Montanuniversität Leoben

Doschek, K., Montanuniversität Leoben

Raupenstrauch, H., Montanuniversität Leoben

Originally presented on: 11/7/2013 10:10:00 - 10:35:00

Blast furnace slag is known as by-product from the production of hot metal at integrated steel and iron mills. Approximately 300 kg of this liquid slag with a temperature of about 1500 °C is tapped together with each ton hot metal. This means a worldwide annual production of about 400 million tons of blast furnace slag. Further processed, water granulated blast furnace slag can be produced, if the liquid slag is cooled down rapidly for solidification. State of the art is a closed system which uses lots of water for rapid cooling and solidification to produce a granulated blast furnace slag which has an amorphous structure. This product normally is dried and grinded for use as a binder in the cement production due to its latent hydraulic properties. The disadvantage of wet systems is on the one hand, the recooling of the circulating water and on the other hand the need for drying of the granulated blast furnace slag, due to the residual moisture up to 20 mass percent. Furthermore the sensible heat of about 1.5 GJ/t of slag is lost. [1]

Therefore a new lab rig has been installed by Siemens VAI in cooperation with Montanuniversität Leoben at the Chair of Thermal Processing Technology for research and development. In this project there are the co-partners voestalpine Stahl GmbH, ThyssenKrupp Steel Europe AG and FEhS-Institut für Baustoff-Forschung e.V. This project is funded by the German Federal Ministry of Economics and Technology (BMWi).

Figure 1: Lab rig at the University of Leoben, Chair of Thermal Processing Technology (Picture: Siemens VAI)

The lab rig is based on the ?Rotating Cup? principle. The liquid slag will be atomized by a cup and small particles spread away due to the occurring forces. First tests involving the ?Rotating Cup? principle were performed in Redcar at British Steel in the 90's. Further investigations took place at Vitkovice in the Czech Republic between 2002 and 2004. These tests were part of the ZeroWaste project of Siemens VAI. Both research activities did not focus on heat recovery due to the lack of political pressure at this time. It could be shown that dry granulation for producing glassy blast furnace slag is also feasible with the ?Rotating Cup?. [1]

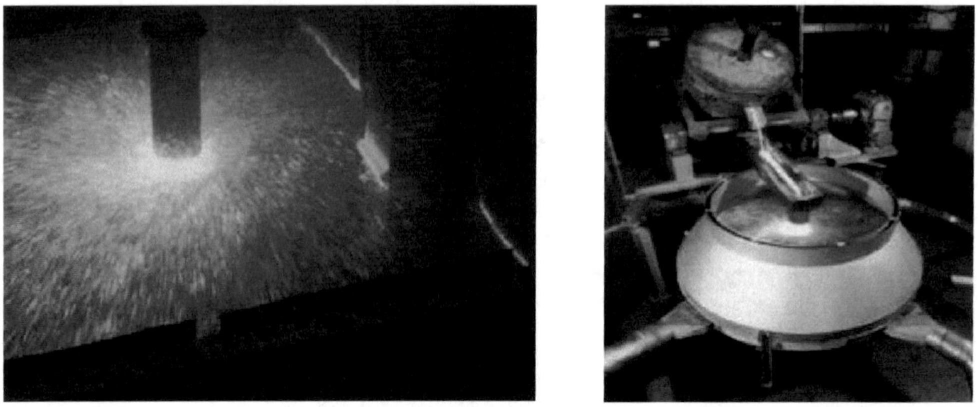

Figure 2: Trials at Redcar (left) and Vitkovice (right) [1]

The new lab rig at the University of Leoben has been designed on basis of the results of the lab tests at FEhS-Institut and the modelling of Siemens VAI to find the best operating parameters for producing high quality granulated blast furnace slag and highest possible air outlet temperatures. Therefore about 300 kg of slag is molten in the ?Flash-Reactor? for each trial at the University's workshop. The liquid slag is tapped into special slag pots which are lined with refractory to prevent heat losses. Afterwards the slag pot is manipulated into a tilting unit, installed close to the granulator. The liquid slag feed into the granulator can be measured and controlled via load cells. The liquid slag will be delivered to the centre of the granulator passing a slag runner and a vertical refractory pipe. After atomizing the slag droplets cool down rapidly during their flight towards the granulator wall once spread away by the fast spinning cup. The surface of the particle has to solidify during this short flight time before falling into a developed fluidized bed made by granulated particles. [1]

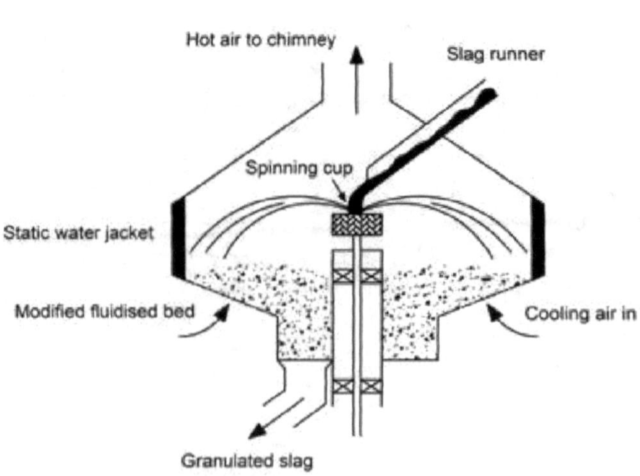

Figure 3: ?Rotating Cup? principle [1]

Furthermore the granulator wall is indirect cooled by a water jacket. Air will be introduced at the bottom of the device. A special air distribution insures a proper cooling of the particles, an agitated bed and no agglomeration. So a maximum heat exchange rate can be reached. Hot off gas can be added to raise the input air temperature for the granulation process. Hence a recirculation system can be simulated which should be used in industrial application. Flow straighteners at the top hinder any carryover of the particles, where the hot air leaves the granulator. [1]

Afterwards the air is cooled down by a spray tower before leaving the lab rig passing a droplet separator and the suction fan. The granulated slag particles can be discharged at the bottom of the granulator after each trial.

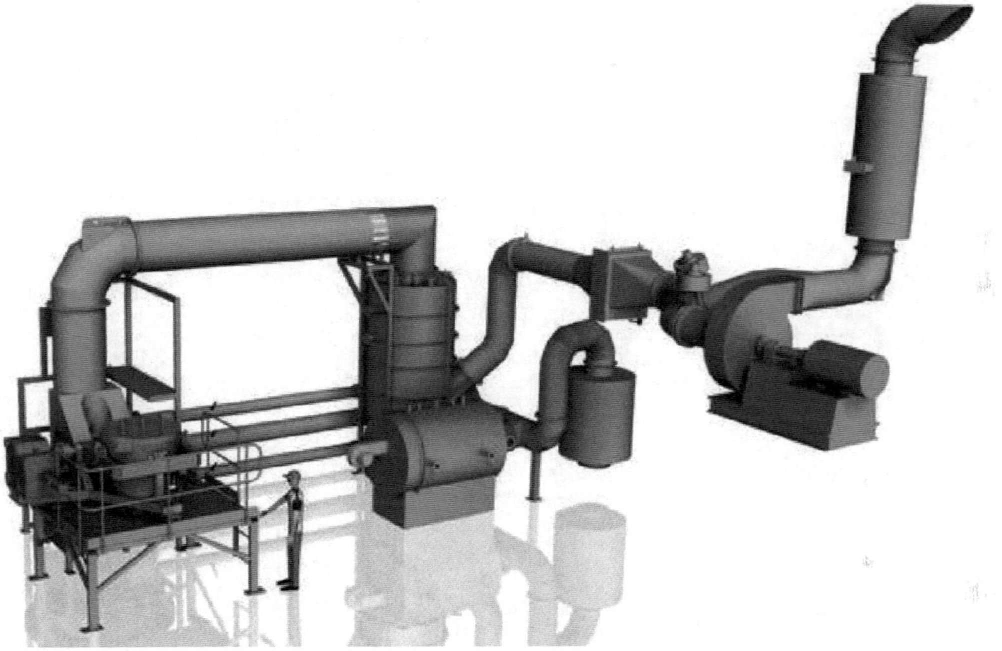

Figure 4: Isometric drawing of the lab rig

The whole lab rig is fully automated. More than 90 sensors are recording the process data like slag mass flow, different pressures, temperatures and other measurands. This helps analyzing each trial to find the best operating parameters for getting a proper product while having a maximum air outlet temperature. The main challenge considers ensuring rapid slag freezing to reach the transformation point of the slag very fast. Otherwise the product will not get fully glassy but partially crystalline which is unsuitable for further use in the cement industry. Cooling with air is much more difficult due to a substantially lower heat capacity compared with water used in conventional devices. Furthermore the air heats up while streaming through the agitated bed. So additionally the temperature difference worsens the setting process.

First campaigns showed promising results regarding glass content, grain size distribution and particle shape. Also a significant off gas temperature increase could be measured. With the obtained data, process optimization can be done and a scale up for further investigations can be realized in the future. Using dry slag granulation at integrated steel and iron mills will gain access to the sensible heat of the blast furnace slag. The recovered energy can be used for the production of steam or electricity or for other useful preheating processes. An energy potential for recovering 20 MW thermal energy or alternatively of about 6 MW electric power generation from a slag feed of one ton per minute can be calculated. Without the need of an upstream drying system for the blast furnace slag energy and as in consequence CO_2 can be saved. In summary dry granulation of blast furnace slag allows an environmental friendly and sustainable subsequent processing for one of the last big heat recovery potentials at a modern blast furnace. [1]

References

[1] McDonald, I., Long, E., Werner, A. & Most, D. (2010) Dry slag granulation - The environmental friendly way to making cement. In: IOM3 conference on Waste Recovery in Ironmaking and Steelmaking Processes. London: United Kingdom

(642f) Advances in Technology for Producing Alumina From Complex Resources in China
Zhang, Y.,
Originally presented on: 11/7/2013 10:35:00 - 11:00:00

The qualified bauxite ore in China to produce Alumina through Bayer process is limited. However, there are considerable complex resources containing significant amount of alumina in China, such as low grade bauxite, fly ash, red mud and nepheline, which can be used to produce alumina through an appropriate process. Compared with low grade bauxite and fly ash, red mud and nepheline, which contain 5-20% of sodium oxide and potassium oxide, could be used to economically extract alumina only through an alkaline sintering or leaching process, because the sodium oxide and the potassium oxide could be leached as valuable hydroxides. However, an acid leaching or sintering process is theoretically suitable to extract alumina from low grade bauxite and fly ash containing tiny sodium or potassium, meanwhile, the purification of the acid leachate is quite important for the development of new acid technology to produce qualified alumina. The improved technologies based on sintering process, alkaline and acid leaching processes to produce alumina from those complex resources in China, and especially, a Low Pressure Hydro-chemical Process (LPHP) leaching technology will be presented.

(655a) Quality By Design Approach for New Chemical Technology Development

Iliyas, A., Saudi Basic Industries Corporation
Alqahtani, A.,
Originally presented on: 11/7/2013 12:30:00 - 12:55:00

In recent years, Quality by Design (QbD) has gained acceptance in the pharmaceutical industry as an efficient way of designing and controlling processes to meet regulatory requirements. QbD approach seeks to use mechanistic, risk-based, holistic and proactive approach to build-in quality into new product development processes. The key steps involves defining design space, conducting risk assessment on process parameters and identifying critical quality attribute.

Current as well as future technology development in petrochemical industry are expected to face greater challenges including market uncertainty, better process (product) performance amidst higher technical uncertainties and stricter environmental sustainability requirement. In this context, SABIC Technology and Innovation (T&I) has recently incorporated QbD approach into earlier stages of its new technology development projects.

This work describes the steps involved in applying QbD concept to new petrochemical process development using an example of one of SABIC technology development. We provide examples of how mechanistic reaction kinetic model was coupled with statistical DoE to enhance chemical process understanding, support assessment of parameter criticality, and help in defining parametric design space which ensure that targeted catalyst activity, products selectivity and purity are achieved at commercial scale. Future roles of mechanistic models to support QbD application in chemical/petrochemical process are discussed.

(655b) Minimize Losses During Scale Up of Multiphase Stirred Tank Reactors Through Better Understanding of Hydrodynamics

Kasat, G., Tridiagonal Solutions Inc.
Vedapuri, D., Tridiagonal Solutions Inc.
Originally presented on: 11/7/2013 12:55:00 - 13:20:00

Stirred reactors are commonly used for carrying out catalytic hydrogenation in the process industries. The catalytic hydrogenation is a complex process involves four phases gaseous(H_2), organic, aqueous and solid (catalyst). A number of competing mass transfer and kinetic rate processes contribute to the overall process performance. The kinetic parameter stay the same on scale-up however other processes such as mass transfer and mixing intensity are usually lower in pilot/plant scale reactors compared to the laboratory scale reactors. This leads to lower conversion and selectivity on scale-up. In a stirred reactor, better mass transfer and mixing intensities can be achieved by selecting proper combination of agitator system, baffle arrangements & operating speed.

In this study Computational Fluid Dynamics (CFD) based models were used to understand the impact of impeller system design, baffle arrangement & operating speed on the reactor hydrodynamics. Multiphase CFD models were used to evolve various performance parameters such as gas-liquid interfacial area, power dissipation, mixing patterns. Multiple configurations were tested and compared with the lab scale reactor and an un-optimized pilot scale agitator design. An optimized design of impeller system & baffle arrangement showing mass transfer rates comparable with lab scale reactor was selected.

The optimized design showed 75% conversion as against the 27% conversion shown by the un-optimized design. The observed conversion with optimized design was in good comparison with the 80% conversion levels observed in the lab scale reactor. Thus, better understanding of reactor hydrodynamics is required for successful scale-up of multiphase stirred reactor. CFD models can be successfully employed to quickly analyze various scenarios and optimize the reactor design

(655c) Scaling Up Liquid-Liquid Extraction Columns From Pilot Plant Tests

Glatz, D. J., Koch Modular Process Systems, LLC

Originally presented on: 11/7/2013 13:20:00 - 13:45:00

Scaling Up Liquid-Liquid Extraction Columns from Pilot Plant Tests

Distillation columns can often be designed without pilot plant testing with a high assurance that the required performance can be achieved in the production columns. However, experience has proven that the most prudent way to design a liquid-liquid (LLE) extraction column is by first testing in a pilot plant column and then using proven scale-up techniques to size the production column correctly. As the diameter of an LLE column increases, the tendency for back mixing to occur also increases and this results in lower efficiency. Thus, the key for scale-up is to understand how the efficiency changes with diameter and to factor this into the design of the production column. This paper will cover the scale up techniques for two agitated extraction columns; the SCHEIBEL® Column with rotating impellors and the KARR® Column with a reciprocating plate stack. Over years of experience with many different LLE systems, proven scale up correlations for these two columns have been developed and used for the design of production columns with performance guarantees. A case study for each column will be presented including the data generated in the pilot plant and then the calculation for sizing the production columns.

(655d) Challenges of Planning and Implementing Biofuels Pilot Plant Projects

Lucero, A. J., Southern Research Institute

Originally presented on: 11/7/2013 13:45:00 - 14:10:00

The Energy Technology Development Center at Southern Research Institute is a unique laboratory that provides technology developers a platform to implement, demonstrate, and validate their pilot plant research projects. Southern Research engineers, operators, and analytical chemists are experienced in all phases of executing pilot plant projects and work with clients to optimize and commercialize their advanced energy technologies, integrate them with value-added technologies, and collaborate with new partners and clients. This presentation will focus on lessons learned planning and implementing multiple biofuels pilot plant projects over the past four years.

(655e) A New Model for the Sizing of Needle Type Control Valves in the Laminar, Transitional and Turbulent Flow Regimes

Martin, P. F.,

Girouard, G.,

Originally presented on: 11/7/2013 14:10:00 - 14:35:00

Control valves for low flows of liquids, even of low viscosity, must be sized to take into account laminar or transitional flow in the valve trim.

We have found that the equation below allows the accurate sizing of control valves with seat-guided type trims for laminar, transitional and turbulent flow:

$$C_{vt}\ required = Q\sqrt{\frac{S.G.}{\Delta P}} \cdot \sqrt{1 + \frac{\mu\, a_v{}^2}{Q \cdot S.G.}}$$

A single new parameter QUOTE a_v a_v is used to adjust the sizing of families of valve trims of similar geometry, for viscosity dependence in the laminar and transitional regions.

This new parameter has a value of approximately 0.14 for seat-guided annular needle-type trims such as those used in Badger Research Control Valves and for Jordan LowFlow control valves for trims of C_V. 0.06 and smaller. A single Samson valve of similar trim type was tested and found to have an QUOTE a_v a_v of 0.18, which we suggest to use for valves of similar design for conservative trim sizing in the absence of other data. We found QUOTE a_v a_v to be approximately constant with varying stem position for trims of this geometry.

For seat-guided vee-groove control valve trims manufactured by Emerson (Baumann), the value of QUOTE a_v a_v is approximately 0.06 at full open, and tends to smaller values as the valve is closed.

This equation can be solved to be explicit in Q and tends to the normal turbulent flow Cv sizing equation when the product QUOTE $\mu a_v{}^2$ $\mu a_v{}^2$ is sufficiently small.

(655f) Enabling Hydroformylation in Micro-Emulsion Systems: Long-Term Performance of a Continuously Operated Mini-Plant

Müller, D., Berlin University of Technology

Esche, E., Berlin University of Technology

Hamerla, T., Berlin Institute of Technology

Rost, A., Berlin Institute of Technology

Kasaka, Y., Berlin Institute of Technology

Schomäcker, R., Berlin University of Technology

Wozny, G., Berlin University of Technology

Originally presented on: 11/7/2013 14:35:00 - 15:00:00

In the Collaborative Research Center SFB/TR 63 InPROMPT, novel process concepts for substitutable base chemicals in multiphase systems are developed. Hereby, next generation liquid-liquid processes based on innovative solvent concepts are currently under research. One of these concepts is concerned with the hydroformylation of long-chained alkenes in micro emulsions. Hydroformylation has been established as a standard process for the production of short-chained aldehydes from alkenes. Its application to higher alkenes (longer than C_{12}) in a biphasic system with a rhodium catalyst, on the other hand, has not yet been established. To speed-up and aid in the process development within the collaborative research center, a mini-plant has been built at Berlin University of Technology (Technische Universität Berlin, TU Berlin).

The investigated process concept aims at reaching two goals: on the one hand, two normally immiscible liquids are to be mixed, while on the other hand a near to perfect separation of reactants is to be achieved. A promising approach lies in the application of a non-ionic surfactant. By creating a micro-emulsion, the surfactant enables the water soluble rhodium-ligand-catalyst required for the reaction to be brought into contact with the alkene [1]. The reaction is started by injecting syngas (H_2&CO) into the system. After the reaction, the miscibility gap between the hydrophilic catalyst solution and hydrophobic alkene/aldehyde mixture is exploited in order to recycle the valuable rhodium catalyst and separate the almost pure organic product phase. To test the viability of this concept, a long-term test-run in the mini-plant at TU Berlin has been performed.

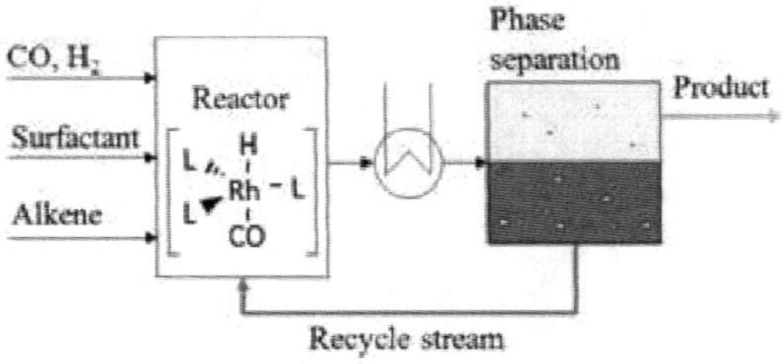

Figure 1: Process concept for the hydroformylation of long-chained aldehydes in micro-emulsions [2].

The mini-plant was operated continuously for 100 hours supported by an operating team of 14 people in four shifts. The applied substances were 1-dodecene (educt), the non-ionic surfactant Marlipal 24/70 (CAS: 68439-50-9), a water soluble rhodium-based catalyst (CAS: 14874-82-9), and the water soluble ligand

Sulfoxantphos (sulfonated form of Xantphos, CAS: 161265-03-8). The desired product is n-tridecanal. The crucial aspects of the concept, with regards to technical and economic feasibility, are the reaction as well as the separation step. The operating conditions for both steps were determined in previous analyses [2 - 4].

During mini-plant operation different reactor residence times, such as 45 and 120 minutes, were tested. In both cases the reactor was operated at 40 bar and 95°C. Batch experiments at similar reaction conditions showed a yield of 2,54% in 30 minutes and 4% in 60 minutes. These results equal those during the start-up of the mini-plant. During continuous operation, an average yield of 1-dodecene to n-tridecanal of about 20% was achieved for a reactor residence time of 45 minutes. In the latter case, where lower feed and recycle streams were applied to increase the reactor residence time to 120 minutes, a yield of about 31% was reached.

The second step of the process is the separation of the catalyst from the product. Here, the phase separation behavior of H_2O-oil-nonionic surfactant systems is exploited [5, 6]. Due to kinetic reasons, the three phase state is desired for the continuous process [2, 3]. Here, a separation into a catalyst-rich, a surfactant rich, and a product rich phase takes place. The lower two phases, catalyst and surfactant, are recycled. The main challenge during plant operation is the shift of the feasible operating points over time due to increasing product concentration inside the system. For the decanter, this means that the temperature needs to be regulated depending on the concentration of product inside the system. During plant operation, a concentration of up to 77 wt.-% of 1-dodecene in the oil and 11 wt.-% in the water phase could successfully be realized in the correct temperature region. Compared to test-tube results, where roughly 90 wt.-% of 1-dodecene remains in the oil phase, this performance is highly satisfactory.

Next to these successful results regarding the reaction and separation step, the long-term operation revealed noteworthy effects. Among these is fractioning of the surfactant during the phase separation step. Thus, another undesired shift of the optimal operating point in the decanter is carried out. As counter measures, surfactant rectification as well as continuous surfactant feeding during plant operation are taken into consideration for consecutive plant runs.

In this contribution selected results from the test run are presented, whereby merits and opportunities for the chemical industry as well as open challenges regarding the process concept are discussed.

Acknowledgment

This work is part of the Collaborative Research Center "Integrated Chemical Processes in Liquid Multiphase Systems" coordinated by the Technische Universität Berlin. Financial support by the German Research Foundation (Deutsche Forschungsgemeinschaft, DFG) is gratefully acknowledged (TRR 63). Furthermore, the authors gratefully acknowledge the support of the company Umicore for sponsoring the rhodium catalyst ?Acetylacetonatodicarbonylrhodium(I) (CAS: 14874-82-9)? used in the described experiments.

References

[1] Kupka, J. A. (2006) Hydroformylierung von 1-Octen in Mikroemulsion. Ph.D. Thesis, Technische Universität Braunschweig.

[2] Müller, D.; Esche, E.; Müller, M.; Wozny, G. (2012) Development of a Short-Cut Model for Three-phase Liquid Separation Dynamics for a Hydroformylation Mini-Plant, Presentation at the AIChE 2012, Pittsburgh, USA.

[3] Müller, M.; Kasaka, Y.; Müller, D.; Schomäcker, R.; Wozny, G. (2013) Process Design for the

Separa-tion of Three Liquid Phases for a Continuous Hydroformylation Process in a Miniplant Scale, Ind. Eng. Chem. Res. DOI: 10.1021/ie302487m.

[4] Hamerla, T.; Rost, A.; Kasaka, Y.; Schomäcker, R. (2012) Hydroformylation of 1-dodecene with water-soluble rhodium catalysts with bidentate ligands in multiphase systems, ChemCatChem. DOI: 10.1002/cctc.201200744.

[5] Sottmann, T.; Stubenrauch, C. (2009) Phase Behaviour, Interfacial Tension and Microstructure of Microemulsions. In Microemulsions: Background, New Concepts, Applications, Perspectives; Wiley-Blackwell: Chichester, U.K., DOI: 10.1002/9781444305524.ch1.

[6] Kahlweit, M.; Lessner, R.; Strey, R. (1983) Influence of the properties of the oil and the surfactant on the phase behavior of systems of the type H2O-oil-nonionic surfactant. J. Phys. Chem., 87, 5032-5040.

(758a) Development and Implementation of a Process to Control Off-Gassing

Lee, A. G., Bristol-Myers Squibb
Lo, E. T., Bristol-Myers Squibb Company
Natalie, K., Bristol-Myers Squibb Co,
Vu, T., Bristol-Myers Squibb Company
Rossano, L., Bristol-Myers Squibb Company
Originally presented on: 11/7/2013 15:15:00 - 15:35:00

The use of the *tert*-butyloxycarbonyl (BOC) moiety is a highly effective protection strategy for amines in the synthesis of small molecule active pharmaceutical ingredients. While evaluating these chemistries, the evolution rate of carbon dioxide and other volatile byproducts from these reactions is a major consideration for scale up. In our development of a BOC protection reaction to produce a pharmaceutical intermediate, the physical property of a solid intermediate severely hindered mixing in the vessel. As the reaction progressed, changes to the solid's physical property resulted in rapid uncontrolled off-gassing. We discuss the development of a process using acid and base charge equivalents, charge rate and pH to address the solid intermediate formation and gas evolution. We scaled up the process in the kilo plant to demonstrate a range of process parameters which yielded good control over gas evolution while maintaining acceptable quality and yield.

(758b) Chemical Engineering Strategies to Directly Deliver Process Properties Across Development and Manufacturing Scales

Glennon, B., UCD
Carr, M., APC Ltd
Whelan, J., University College Dublin
Barrett, M., University College Dublin
Originally presented on: 11/7/2013 15:35:00 - 15:55:00

In recent years there has been a growing necessity to reduce the time, cost and risk associated to the development and scale-up of pharmaceutical processes. The increasing pressures to direct deliver critical quality attributes (CQAs) and process performance criterion (cycle times, capacities etc) across manufacturing and supply scales compounding this challenge. As a result of this process engineering strategies, to enable streamlined, reliable and robust scale-up, has become increasing important.

Within this presentation novel process engineering strategies and technologies, underpinned by process modelling, will be presented that enable the delivery of process characteristics (i.e. particle size, shapes, PSDs, flowability, reaction purities, process kinetics etc). Applications of advanced process control, process simulation, kinetic assessments, particle engineering strategies, processing technologies (batch and continuous) will be discussed via numerous development and scale-up case studies.

(758c) The Chemical Engineer Strikes Back â€“ A New Hope for Commercial Process Development
Rothstein, S.,
Originally presented on: 11/7/2013 15:55:00 - 16:15:00

The process challenges of optimizing and scaling up the reaction to make 2,6-diaminopyrazine-1-oxide (DAPO), a precursor to a highly promising insensitive energetic material, will be discussed. The original process featured a precisely controlled dual addition of a liquid base and a solid reagent to limit the formation of more than 30% impurity and minimize starting material decomposition. These conditions limited this labor intensive reaction to the laboratory scale and produced DAPO with inconsistent yield and purity. Our understanding of the chemical reactivity led to improved reaction conditions featuring liquid doses rather than a solid dose, thus allowing for the rate of the reaction to be controlled by automated dosing. Using a kinetic model built with DynoChem® software, a robust process was optimized to improve the yield of DAPO and significantly reduce impurity formation. The reaction was scaled up using the optimized conditions and provides a potential commercial process for DAPO.

(758d) Reaction Engineering for a Sulfonamide Intermediate

Marek, J. C., AbbVie, Inc.
Zhao, Y., AbbVie, Inc.
Hill, D. R., AbbVie, Inc.
Originally presented on: 11/7/2013 16:15:00 - 16:35:00

Reaction engineering of a sulfonamide synthesis from early through mid and late stage development is discussed, with emphasis on consideration of reaction pathways. Reaction of sulfonyl chloride with ammonia in acetone solvent eliminates the use of aqueous ammonium hydroxide at extreme cryogenic temperatures (-60 °C), or for solid ammonium carbamate to achieve the amidation. The order of addition in the semi-batch amidation reaction was chosen to minimize use of excess equivalents of amidation reagent. In the revised process, the ammonia / acetone solution is charged into a solution of sulfonyl chloride, maintained at sub-ambient temperature (NMT -10 °C). Hence, both temperature and reagent equivalents are used to control aniline sulfonamide impurity levels. Formerly, the sulfonyl chloride was added to an excess of ammonium carbamate reagent at sub-ambient temperatures, and was promptly quenched to avoid over-reaction. The revised process affords robustness in design space including less potential for over reaction to aniline impurity.

(758e) Scale-Up Oscillatory Helical-Baffled Reactor for Multiphase Reactions

Phan, A. N., Newcastle University
Harvey, A. P., Newcastle University
Originally presented on: 11/7/2013 16:35:00 - 16:55:00

Scale-up oscillatory helical-baffled reactor for multiphase reactions

Anh N. Phan and Adam P. Harvey

School of Chemical Engineering & Advanced Materials (CEAM), Newcastle University, NE1 7RU, UK

Corresponding to anh.phan@ncl.ac.uk

Mesoscale oscillatory helical-baffled reactor (MOHB), a helical wire embedded inside a smooth tube without modification to the tube, was constructed and characterised at very low net flow rates (net flow Reynolds numbers, $Re_n \leq 10$) over a wide range of oscillation conditions. The fluid mixing is controlled by the oscillatory flow and is thus independent of net flow. The flow becomes chaotic at high oscillatory Reynolds numbers and the secondary flow (swirling) becomes active, which imposes an axial velocity component. The coupled effect of the vortex flow and the swirling flow in the presence of helical baffles augments the significant radial mixing and greatly broadens the operating/design window of these reactors. Plug flow performance can be achieved at a ratio of oscillatory Reynolds number to net flow Reynolds number of approximately 10-250 and a helical turn to tube diameter ratio in a range of 0.2-0.6. The highest degree of plug flow was achieved at Strouhal number $0.2 < St \leq 0.1$ and $Re_o = 400\text{-}700$ with the number of tanks in series equal to the number of helical turns. At a scale of 5mm inner diameter, the MOHB is considered to be effective and robust in scale-down production of high added value products, decreasing the reagents requirements and waste.

Is there a correlation between this mesoscale and a conventional scale? We scaled the mesoscale reactor up to 13 times and 130 times in volume by maintaining geometrical parameters (opened cross-sectional area and helical turn to tube diameter ratio) and oscillatory conditions.Â It was found that the same mixing behaviour in terms of residence time distribution profiles of tracers was obtained in the large scales as in the mesoscale.

Keywords: scale-up, helical-baffled reactor, oscillatory flow, plug flow

(758f) Process Scale-Up From Drying of a Single Drop to Production Scale Spray Dryer Using Levitated Single Droplet Analysis (DKA) and Computation Fluid Dynamics (CFD)

Walter, M., GEA Process Engineering Inc.

Originally presented on: 11/7/2013 16:55:00 - 17:15:00

Process Scale-Up from Drying of a Single Drop to Production Scale Spray Dryer Using Levitated Single Droplet Analysis (DKA) and Computation Fluid Dynamics (CFD)

Since its recent introduction, the single droplet analyzer (DKA - Digital Kinetics Analyzer) and Computational Fluid Dynamics (CFD) has allowed optimization of spray dryer design. Prior to utilizing DKA combined with CFD, one would have to theorize how a powder would behave during drying under various conditions and design the spray dryer accordingly. New Computational Fluid Dynamics modeling power combined with a novel means to examine in detail how a single drop of feed dries has allowed spray dryers to be more efficient, safer, smaller, of higher capacity, and with fewer operational headaches than in the past.

In addition to improvements in spray dryer design, DKA allows us to examine how individual particles dry and how different drying parameters affect the dried product. This allows the spray dryer to produce a ?custom particle? through adjustment of spray drying parameters determined by DKA. For example, some might want a strong particle not easily damaged during transport; others might require a lightweight powder that dissolves easily. Use of DKA can determine if a given feed will dry into hollow spheres, shriveled particles, ?doughnut? shapes, etc. while providing insight into design of drying equipment and process parameters for attaining a certain type of particle.

(758g) An Integrated Management System to Optimize Process Operations Within the Pharmaceutical Industry

Drayton, R. A., THS, Inc
Gloor, G. J., Imperial College London
Grills, J.,
Originally presented on: 11/7/2013 17:15:00 - 17:35:00

One of the major challenges in developing a pharmaceutical process is the need to supply material of a quality representative to the final product, and at the same time, develop a safe and scalable process that can evolve into the final manufacturing process. In order to overcome this challenge, requires careful integration of process chemistry development, analytical techniques and process engineering of the chemical synthetic route. Historically, process safety has been regarded as a manufacturing concern, leaving the collection and analysis of process safety information, such as thermal stability and thermo-chemistry to the latter stages of process development. Since, other constraints such as product impurity was developed earlier in the process, this introduced the need for engineering solutions to process safety problems later in the process. This could have been avoided by straightforward changes in process chemistry, had these problems been identified and determined earlier in development. In addition, the process safety information that is obtained from physical experiment to address the process safety problems, can also provide valuable insight into, *inter alia*, process robustness and optimum operating conditions. Thus, the opportunity to leverage this information is lost by separating process safety from process chemistry and regarding it as an independent compliance function.

It is proposed here, to integrate the activities generally associated with process safety, such as thermal stability testing, reaction engineering and safety management into the development process for pharmaceuticals and fine chemicals. This will enable the data generated by the process safety function to be part of the dossier of process information that can be utilized by chemists and engineers in developing the "operating envelope". In addition, companies can avoid the delay and expense due to process changes that result from safety recommendations at latter stages in the process, and which conflicts with the process declarations in regulatory filings. This paper presents a novel integrated management system that can optimize process operations within the pharmaceutical industry.

AUTHOR INDEX

AUTHOR INDEX

AUTHOR INDEX

AUTHOR INDEX

AUTHOR INDEX

AUTHOR INDEX

AUTHOR INDEX